IDS and IPS with Snort 3

Get up and running with Snort 3 and discover effective
solutions to your security issues

Ashley Thomas

IDS and IPS with Snort 3

Copyright © 2024 Packt Publishing

Group Product Manager: Pavan Ramchandani

Publishing Product Manager: Neha Sharma

Book Project Manager: Ashwin Dinesh Kharwa

Senior Editor: Apramit Bhattacharya

Technical Editor: Irfa Ansari

Copy Editor: Safis Editing

Proofreader: Apramit Bhattacharya

Indexer: Subalakshmi Govindhan

Production Designer: Vijay Kamble

DevRel Marketing Coordinator: Kamalpreet Kaur Sahni

First published: September 2024

Production reference: 1040924

Published by Packt Publishing Ltd.

Grosvenor House

11 St Paul's Square

Birmingham

B3 1RB, UK

ISBN 978-1-80056-616-3

www.packtpub.com

To my son, Rohan.

Contributors

About the author

Ashley Thomas is a security researcher at Dell SecureWorks and a member of the Counter Threat Unit team. Before this role, he was instrumental in building the iSensor, a proprietary network intrusion prevention system. Ashley has a master's in computer networking from North Carolina State University, and he also holds several other certifications, including CISSP, GCIA, GREM, GCLD, and GWEB. He has authored several papers on intrusion detection and holds several patents in this field.

About the reviewers

Ron Cowen has been in the network security industry for over a decade, spanning roles at AT&T, Juniper Networks, and his current position as a senior systems engineer for Palo Alto Networks. He is based in Seattle, Washington.

I'd like to acknowledge and thank all of those who have supported, and those who continue to support, my growth as a network security professional, as well as my wife and our two daughters

Wayne Burke, VP for `Cyber2labs.com`, is internationally recognized for his commitment, achievements, and contributions to the IT security industry. He currently specializes in many offensive and defensive AI technologies for robotics such as drones, building and managing new high-tech security tools, custom hardware solutions for bio-medical products, digital forensics, penetration testing, and mobile security, and radio-frequency SDRs. Wayne and his team have delivered security assessments, penetration test assignments, and customized training for international corporations and many government agencies, such as EPA, FAA, DOJ, DOE, DOD, the Air Force, the Army, the Navy, the Marines, CIA, FBI, NSA, and many more US government bodies.

Table of Contents

2

Part 2: Snort 3 – The New Horizon

3

4

5

Configuring Snort 3   53

Part 3: Snort 3 Packet Analysis

6

Data Acquisition   81

7

Packet Decoding   95

8

Inspectors   109

9

Stream Inspectors   119

10

HTTP Inspector   143

11

DCE/RPC Inspectors   155

12

IP Reputation   165

Part 4: Rules and Alerting

13

Rules   177

Preface

Snort is recognized as the industry standard for intrusion detection and prevention systems; Snort 3 is the latest version of the software and includes significant changes to its functionality and features. This book will introduce you to IDS/IPS systems and the Snort IDS/IPS system. It will provide you with details on the latest version of Snort, Snort 3, and familiarize you with its workings and its configuration.

Who this book is for

This book is for anyone who wants to learn about Snort 3. If you are a beginner to the world of Snort, or if you are someone who has used Snort and would like to learn about its latest version, this book is for you.

We expect that network administrators, security administrators, security consultants, and other security professionals will find this book useful. Those using other intrusion detection systems (IDS) will also gain from this book as it covers the basic inner workings of any IDS. Although there are no prerequisites, basic familiarity with Linux systems and knowledge of basic network packet analysis will be very helpful.

What this book covers

Chapter 1, *Introduction to Intrusion Detection and Prevention*, discusses a defense-in-depth strategy and the role of various security tools, including IDS/IPS.

Chapter 2, *The History and Evolution of Snort*, explores the evolution of Snort from its original version to its current state. We will look at the key features of Snort and when they were incorporated into the system.

Chapter 3, *Snort 3 – System Architecture and Functionality*, explores the design goals, the main components, and the system architecture of Snort 3. The chapter provides you with a high-level idea of how network traffic gets analyzed by the Snort system.

Chapter 4, *Installing Snort 3*, shows you how to install the Snort 3 system. The chapter describes the step-by-step installation process of Snort 3 on two different operating systems.

Chapter 5, *Configuring Snort 3*, explains how to configure the Snort 3 system. It discusses how a user can configure the Snort 3 system and the various modules, using command-line arguments as well as configuration files.

Chapter 6, Data Acquisition, delves into the data acquisition layer and its role in the delivery and transmission of network packets to and from Snort.

Chapter 7, Packet Decoding, reinforces the idea that an analysis of network traffic begins with packet decoding. This chapter explains the process of packet decoding and discusses how the packet decoding module is structured, what the important data structures are, and how the module ties to the rest of the Snort system.

Chapter 8, Inspectors, discusses inspectors, which are considered the backbone of Snort 3 from a functionality perspective. From an evolution standpoint, the inspectors replaced the preprocessor module in Snort 2. This chapter discusses the role and functionality of the Inspector modules.

Chapter 9, Stream Inspectors, discusses the stateful analysis capability of Snort 3. The chapter also explains important terms such as *flows*, *sessions*, and *streams*, which are relevant to how Snort performs stateful analysis.

Chapter 10, HTTP Inspector, explores HTTP, which is one of the most prevalent protocols used over the internet. This chapter discusses the HTTP inspector and how it enables the detection of malicious attacks over the HTTP protocol.

Chapter 11, DCE/RPC Inspectors, discusses the DCE/RPC inspectors and their overview, dependencies, relevant rule options, and configurations.

Chapter 12, IP Reputation, shows you how the IP reputation inspector module works, its configuration, and its importance.

Chapter 13, Rules, discusses how Snort rules work, its structure, and some important points to keep in mind while developing Snort rules. The use of Snort rules allows a Snort user to specify what constitutes malicious traffic.

Chapter 14, Alert Subsystem, delves into the *alert* subsystem of Snort. We will discuss the various alert modules and how they are configured.

Chapter 15, OpenAppID, discusses the OpenAppID feature, the relevant inspector modules, and their configuration.

Chapter 16, Miscellaneous Topics on Snort 3, discusses a handful of miscellaneous topics related to Snort 3. We will explore how to go about troubleshooting and/or debugging Snort, Snort 2 to Snort 3 migration challenges, and so on.

To get the most out of this book

You are expected to know the basics of computer networking, networking protocols, and traffic analysis. Familiarity with network traffic analysis tools such as Wireshark and/or tcpdump will be useful. Familiarity with Linux operating systems is also expected.

Software/hardware covered in the book	Operating system requirements
Snort 3	Linux

If you are using the digital version of this book, we advise you to type the code yourself or access the code from the book's GitHub repository (a link is available in the next section). Doing so will help you avoid any potential errors related to the copying and pasting of code.

Download the example code files

You can download the example code files for this book from GitHub at `https://github.com/PacktPublishing/IDS-and-IPS-with-Snort-3.0`. If there's an update to the code, it will be updated in the GitHub repository.

We also have other code bundles from our rich catalog of books and videos available at `https://github.com/PacktPublishing/`. Check them out!

Conventions used

There are a number of text conventions used throughout this book.

`Code in text`: Indicates code words in text, database table names, folder names, filenames, file extensions, pathnames, dummy URLs, user input, and Twitter handles. Here is an example: "Another key component that was included in this release was the IP `defrag` module."

A block of code is set as follows:

```
alert tcp any any -> $HOME_NET [80,8080] (msg:"SQL Injection
Detected"; flow:established,to_server; http_uri; content:"/
wordpress/wp-content/plugins/demo_vul/endpoint.php"; content:"union";
distance:0; http_uri; content:"select"; distance:0; nocase;
content:"from"; distance:0; nocase; sid:123;)
```

When we wish to draw your attention to a particular part of a code block, the relevant lines or items are set in bold:

```
http://acunetix.php.example/wordpress/wp-content/plugins/demo_vul/
endpoint.php?user=-1+union+select+1,2,3,4,5,6,7,8,9,(SELECT+user_
pass+FROM+wp_users+WHERE+ID=1)
```

Any command-line input or output is written as follows:

```
sudo dnf install -y flex bison gcc gcc-c++ make cmake automake
autoconf libtool curl pkgconf
```

Bold: Indicates a new term, an important word, or words that you see on screen. For instance, words in menus or dialog boxes appear in **bold**. Here is an example: "It can be noted that the **Total Length** field is 16 bits."

> **Tips or important notes**
> Appear like this.

Get in touch

Feedback from our readers is always welcome.

General feedback: If you have questions about any aspect of this book, email us at customercare@packtpub.com and mention the book title in the subject of your message.

Errata: Although we have taken every care to ensure the accuracy of our content, mistakes do happen. If you have found a mistake in this book, we would be grateful if you would report this to us. Please visit www.packtpub.com/support/errata and fill in the form.

Piracy: If you come across any illegal copies of our works in any form on the internet, we would be grateful if you would provide us with the location address or website name. Please contact us at copyright@packt.com with a link to the material.

If you are interested in becoming an author: If there is a topic that you have expertise in and you are interested in either writing or contributing to a book, please visit authors.packtpub.com.

Share Your Thoughts

Once you've read *IDS and IPS with Snort 3*, we'd love to hear your thoughts! Scan the QR code below to go straight to the Amazon review page for this book and share your feedback.

https://packt.link/r/1-800-56616-6

Your review is important to us and the tech community and will help us make sure we're delivering excellent quality content.

Download a free PDF copy of this book

Thanks for purchasing this book!

Do you like to read on the go but are unable to carry your print books everywhere?

Is your eBook purchase not compatible with the device of your choice?

Don't worry, now with every Packt book you get a DRM-free PDF version of that book at no cost.

Read anywhere, any place, on any device. Search, copy, and paste code from your favorite technical books directly into your application.

The perks don't stop there, you can get exclusive access to discounts, newsletters, and great free content in your inbox daily

Follow these simple steps to get the benefits:

1. Scan the QR code or visit the link below

https://packt.link/free-ebook/9781800566163

2. Submit your proof of purchase
3. That's it! We'll send your free PDF and other benefits to your email directly

Part 1: The Background

Information security plays a crucial role in the successful operation of any organization. **Intrusion Detection Systems (IDS)** and **Intrusion Prevention Systems (IPS)** have a pivotal role in a defense-in-depth information security strategy. One of the leading open source IDS/IPS systems of our time is Snort.

The first part of the book covers the necessary background information about network security and intrusion detection, the role of information security, and the role of an IDS/IPS system within a defense-in-depth strategy. A brief history of the evolution of Snort to its current state is also provided. With this background, we will start discussing Snort 3 in the second part of the book.

This part has the following chapters:

- *Chapter 1, Introduction to Intrusion Detection and Prevention*
- *Chapter 2, The History and Evolution of Snort*

1

Introduction to Intrusion Detection and Prevention

Information security plays a key role in the successful operation of any organization; it ensures the confidentiality, integrity, and availability of information. **Intrusion detection systems** (IDS) and **intrusion prevention systems** (IPS) play a critical role in the defense-in-depth strategy used in the information security field. Historically, the role of intrusion detection was primarily that of monitoring in order to detect malicious or suspicious activity. Over time, the prevention capability was added in addition to detection, thereby creating IPS. As the nature of computation evolved over time, the nature of threat and attack vectors also evolved. Subsequently, the complexity of analysis and computation required by intrusion detection has also evolved in order to address the threat landscape. This chapter will introduce you to IDS and IPS at a high level. The chapter will cover the following topics:

- The need for information security
- Defense-in-depth strategy
- The role of network IDS and IPS
- Types of intrusion detection
- The state of the art in IDS/IPS
- IDS/IPS metrics
- Evasions and attacks

The need for information security

Software and IT are everywhere, and their adoption is increasing at an ever-increasing speed. Software programming is prevalent in the fields of entertainment, health, education, food, travel, auto, communication, media, and every other field we can think of. As the number of software programs and their features increase, so does the number of software bugs and flaws. A security flaw, glitch, or weakness found in software code that could be exploited by an attacker (threat source) is called a software vulnerability. The number of such vulnerabilities has been increasing drastically year by year, as seen in the following figure.

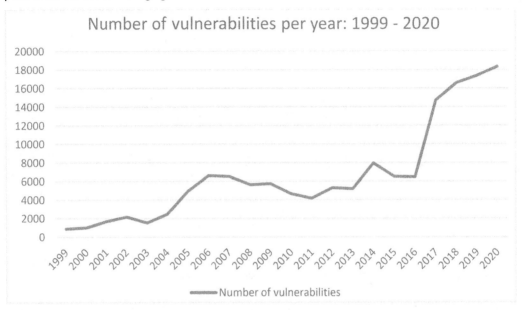

Figure 1.1 – Vulnerabilities trend over the past decades

Threat actors take advantage of such vulnerabilities and cause disruption to the confidentiality, integrity, or availability of the protected system. In certain vulnerabilities, the threat actor makes use of various exploits to deliver, install, and/or execute a malicious program on the system. Such malicious code is known as *malware*.

Malware comes in a variety of forms – viruses, worms, backdoors, trojans, adware, spyware, ransomware, and so on – each with its own characteristics. This malware aims to steal, damage, and/or destroy vulnerable systems – exfiltrating sensitive data or encrypting files and/or disks to make them unusable.

The damage caused by ransomware alone is shown in the following chart:

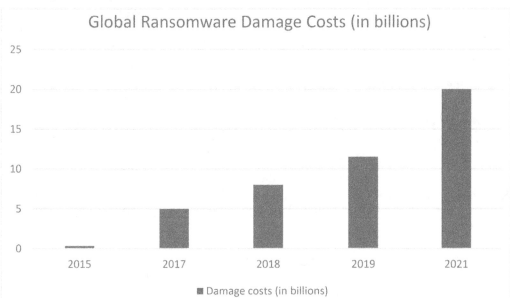

Figure 1.2 – Increasing cost of ransomware-related damage

Typical cyberattacks consist of a set of common phases or stages. Lockheed Martin has created a model called the Cyber Kill Chain to encapsulate these stages (`https://www.lockheedmartin.com/en-us/capabilities/cyber/cyber-kill-chain.html`). The stages are as follows:

1. **Reconnaissance**: This is the phase in which the adversary identifies the target's possible vulnerabilities and weak points. This may involve active scanning of the target network, passive information gathering, social engineering, gathering information from the internet and/or social networks, and so on. This step provides the adversary with sufficient information to proceed with the attack – such as which IP addresses are accessible, what ports are open, what applications are running, and details of the vulnerabilities on each.

2. **Weaponization**: In this stage, the attacker creates a payload (weapon) that exploits the discovered vulnerability and plants malware on the victim's machine.

3. **Delivery**: This is the stage when the attacker delivers the prepared payload, for example, an infected document to the target. A typical delivery mechanism is a phishing email containing a malicious link or an infected PDF document.

4. **Exploitation**: In this stage, the target machine is compromised by the exploit delivered in the previous stage. When the exploit code is executed, the attacker accomplishes their objective, such as remote control of the target machine. Subsequently, having gained a foothold on the victim's machine, the attacker proceeds to the next phases, such as maintaining persistence and exfiltrating data.

5. **Installation**: In the installation phase, various types of malwares are installed on the target machine – ransomware, backdoors, or trojans – based on the plan of the attacker for their purposes.

6. **Command and control**: Once the malware is installed on the target machine, it typically contacts a command and control server. This may be to get additional instructions or commands to be executed on the target machine.

7. **Action**: In this stage, the malware acts on the target as per the commands or instructions from the attacker. This may involve installing additional malware, exfiltrating sensitive data and system information back to the attacker-controlled server, or even performing denial-of-service attacks on any specified targets.

These are the typical stages of a cyberattack. From a security point of view, the earlier the attack is detected, the better. If the defense mechanisms in place can detect and stop an attack at the delivery stage, any compromise can be prevented.

In the next section, let us look at a strategy that aims to ensure the highest chance of a successful defense against attack attempts.

Defense-in-depth strategy

Defense in depth is a strategy for protecting a system against any attack using several independent defense methods. This approach was originally conceived by the National Security Agency. The system that needs to be protected consists of a set of resources and assets, including the network itself. A typical scenario would include web servers, mail servers, DNS infrastructure, WAN and LAN routers, authentication servers, database servers, laptops, and desktops.

As mentioned earlier, a defense-in-depth strategy uses independent and mutually exclusive mechanisms to protect and defend the assets; thus, the chances of detecting an attack are higher than using a single mechanism. It is sufficient for any one of the layers to detect the attack, in order to prevent and thwart it. The several layers of the defense-in-depth strategy are depicted in *Figure 1.3*.

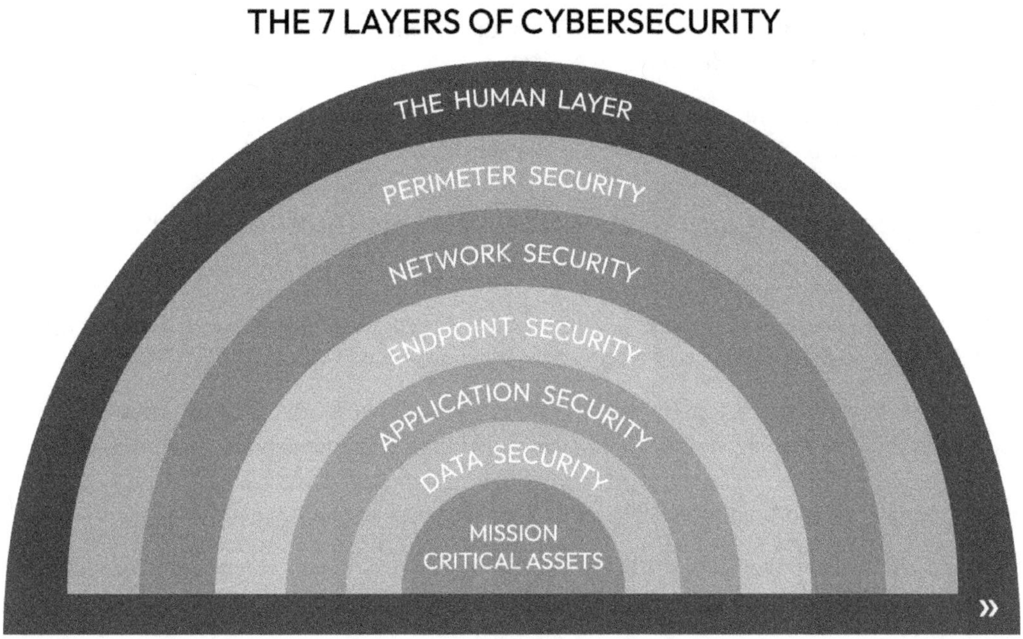

THE 7 LAYERS OF CYBERSECURITY

THE HUMAN LAYER

PERIMETER SECURITY

NETWORK SECURITY

ENDPOINT SECURITY

APPLICATION SECURITY

DATA SECURITY

MISSION
CRITICAL ASSETS

Figure 1.3 – Defense in depth

The defense-in-depth strategy would include security technology, processes, and/or policies at several layers, including network, perimeter, endpoint, application, and data security.

Some of the various layers of the defense-in-depth approach in a typical scenario are discussed in the following subsections.

Firewalls (network and host layers)

Network firewalls filter the network by inspecting traffic that enters or leaves through network boundaries/zones. They enforce user-defined security policies across single or multiple network segments, comparing policies, adding threat modules, and assessing the data packets to prevent unauthorized access. Firewall deployments are precisely placed within the network to inspect and manage traffic flow.

Network firewalls are analogous to doorkeepers. When deployed in the network perimeter, they are typically the outermost layer in the defense-in-depth strategy. However, network firewalls are also deployed within a segregated network to separate various sections and/or departments. Network firewalls perform basic protocol decoding and analysis in order to be able to allow or deny packets and/or connections in or out of the network.

Host-based firewalls are like network firewalls except that they are concerned only with a single host as opposed to a set of hosts in a network.

Network- and host-based firewalls can create logs for every inbound and outbound connection that traverses through them. This can be immensely valuable from a detection point of view.

Intrusion detection and prevention systems (network and host layers)

IDS are analogous to security cameras. They are devices or programs that detect malicious activity against the concerned network or host (network-based or host-based IDS).

For a network-based IDS, the system inspects and analyzes the network traffic and tries to detect malicious activity based on signatures (for known attacks) or anomalous behavior or deviation from standard. The deviation from the standard can either be a statistical deviation (statistical anomaly-based IDS) or a deviation from protocol specifications (protocol anomaly-based IDS).

A host-based IDS will monitor all host artifacts in order to detect malicious activity, including network traffic to or from that host, process details, host-based logs, and files on the host.

IPS are IDS with the additional capability to enforce actions that *prevent* an attack. For example, upon detection of an attack, the IPS may drop the concerned packet or block the entire connection.

Endpoint detection and response (host layer)

Endpoint detection and response (EDR) comprises tools and technology that monitor activity on endpoint hosts and servers in order to detect malicious activity. The activity that is monitored by EDR includes processes, connections (to and from) the host, files created/modified, and registry changes.

Web application firewalls (network and host layers)

Web application firewalls (WAF) are firewalls specifically for web traffic. WAF inspect and analyze web traffic comprehensively. They can analyze both HTTP and HTTPS protocols. In the case of HTTPS, WAF often terminate the SSL sessions to decrypt the traffic, which often involves playing a man-in-the-middle role between the web client and the web server.

Traditional firewalls allow or deny traffic based on OSI layer 3 and 4 headers. Network-based IPS can perform limited application-level analysis. Compared to these, WAF are capable of comprehensive web (HTTP/HTTPS) traffic analysis in order to make the allow versus deny decision.

Some of the commercial companies that offer WAF are Fortinet, Barracuda, and Imperva. ModSecurity is also a widely available option for an open source WAF.

Mail security gateway (network)

A mail security gateway or firewall is another application-level firewall but for email-related protocols. A significant percentage of threats involve emails. In the first half of 2021, 75% of threats were delivered using email. Emails are often used as bait to trap unsuspecting users – by prompting them to open a malicious attachment, or by tempting them to click a malicious link.

Mail security gateways protect users from threats related to email by analyzing and filtering the malicious artifacts from an email. Mail firewalls perform deep inspection of the protocols related to mail, namely SMTP, POP, IMAP, and their encrypted counterparts.

Log management and monitoring (network and host)

Log management and monitoring solutions collect, inspect, and archive log messages and files from a variety of devices in the network. They also enable capabilities such as indexing and searching across the collected logs.

In the next section, let us specifically look at network IDS and IPS and the role that they play in the defense-in-depth strategy.

The role of network IDS and IPS

Network-based IDS and IPS play a significant role in the defense-in-depth strategy for information security. This role is unique when compared with other pieces of the defense-in-depth approach. As the name suggests, the primary role of IDS is detection, whereas IPS adds the extra capability of blocking the attack that it has detected.

The network IDS processes network traffic – analyzes the various protocols that are involved – with the goal of detecting malicious activity in a real-time fashion. The network IDS typically also has the capability to analyze packet captures offline; however, the most common case is to perform the analysis live so as to detect the attack in real time.

In general, the network IDS functionality would include the following:

- **Configuration management**: IDS configuration essentially determines what exact functionality is performed by the IDS, how much memory needs to be allotted, the various parameters for learning for anomaly-based IDS, and the signatures to be analyzed.

- **Packet acquisition module**: This module is responsible for getting the network traffic data (packet data) from the source to the IDS. IDS often use packet capture libraries such as libpcap in order to attain this functionality.

- **Decoder module**: Irrespective of the type of IDS (signature-based or anomaly-based), there needs to be a module that can decode the various network protocols, maintain some state, and make the data available for the rest of the IDS to perform its detection operation.

- **Detection module**: This is the module that performs the detection functionality – whether it is signature matching or detecting an anomaly.

- **Alert and log module**: This module performs the task of generating an alert in the event of attack detection, as well as logging critical log messages regarding the IDS operation.

In the event of detecting an attack, the IDS/IPS generates an alert; these alerts are brought to the attention of a security operator for further action or sent to a central system such as **Security Incident and Event Management (SIEM)** for collection, correlation, and analysis. *Figure 1.4* shows a typical IDS and IPS deployment scenario. It can be noted that the IPS is deployed in an *inline* fashion, whereas the IDS is deployed in an *offline* manner.

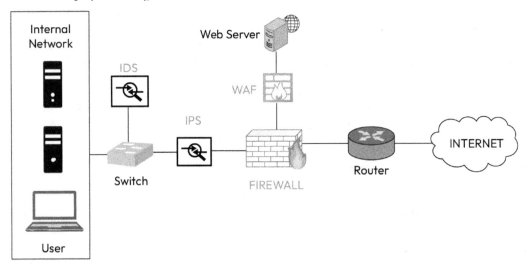

Figure 1.4 – Typical IDS and IPS deployment diagram

Due to the difference in their objectives, the IDS is typically deployed in a passive manner, often analyzing a copy of the network traffic (collected via a SPAN port on a router or firewall). IPS devices, on the other hand, operate in an inline mode – very similar to a firewall – so that they can block the offending packet or connection.

This difference – passive/offline versus inline – in the deployment leads to a key distinction. When the traffic rate increases to a level that the IDS cannot keep up with, it leads to packet drops; it does not affect the operation as it is a copy of the packet that was dropped. However, in the case of an inline operation, when the IPS cannot keep up with the rate of traffic leading to packet drops, it affects the network throughput and becomes a performance bottleneck. Therefore, there is increased demand on the IPS to have faster packet processing than for an IDS.

There is yet another key difference between the IDS and IPS, namely the consequence of a false positive. A false positive is when the IDS or IPS detects a benign packet or connection as malicious. For an IDS, this will result in a false positive alert being generated. This will result in an unnecessary alert and analysis. However, for an IPS that blocks packets and connections when an alert is generated, this will result in the interruption of a normal or benign connection, resulting in user dissatisfaction.

Due to these key differences, IDS and IPS devices are often configured very differently – one giving priority to detection (IDS) and the other giving priority to performance as well as detection (IPS).

In the next section, we will discuss how the IDS and IPS are categorized based on how the *detection* is done.

Types of intrusion detection

Intrusion detection approaches are classified into the following based on how malicious activity is detected. The most common approaches are signature-based, anomaly-based, and hybrid. Let us discuss each of these approaches.

Signature-based intrusion detection

The signature-based approach uses predefined signatures in order to detect known threats. When an attack is initiated that matches one of these signatures, a predefined action (for example, generate an alert) is taken.

This is the most common approach for intrusion detection, especially in commercial solutions. Open source IDS/IPS – such as Snort and Suricata – are essentially signature-based. Signature-based systems are very good and proven to detect known attacks with very good accuracy and efficiency. As opposed to anomaly detection techniques, the signature-based IDS does not require any training or learning phase. The most important disadvantage of this approach is the inability to detect unknown attacks. Due to this reason, this approach requires constant (almost daily) updates to the signature set so that it can detect new threats that appear daily.

A simplified block diagram of a signature-based IDS is shown in *Figure 1.5*.

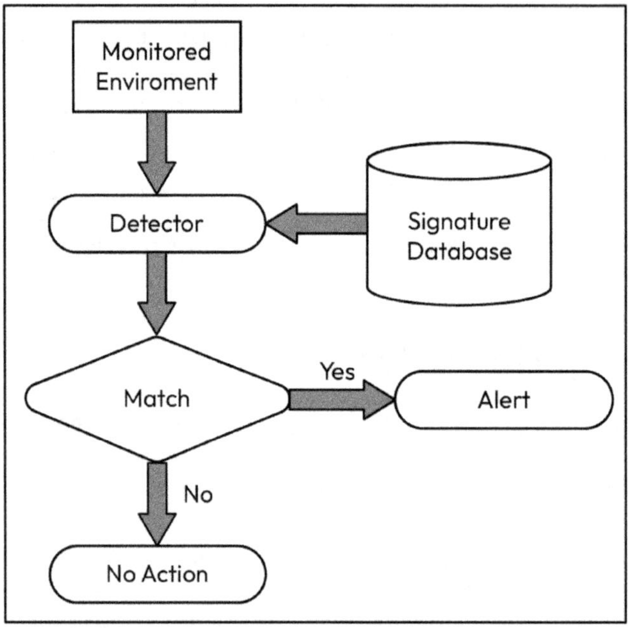

Figure 1.5 – Block diagram of a typical signature-based IDS

The input from the monitored environment (for example, packets from a monitored network) is processed and matched against a set of signatures; if there is a match, the system generates an alert. The quality of the system clearly depends on the quality of the signatures, and therefore maintaining and keeping the signatures updated is one of the main challenges of the system. The race between the attacker, who tries to create an exploit for a newly known vulnerability, and the defender (security operator), who attempts to create a signature that detects attacks against that vulnerability, is often a race against time.

Here is an example of an IDS (Snort) signature:

```
alert tcp any any -> $HOME_NET [80,8080] (msg:"SQL Injection
Detected"; flow:established,to_server; http_uri; content:"/wordpress/
wp-content/plugins/demo_vul/endpoint.php"; content:"union",distance
0; content:"select",distance 0,nocase; content:"from", distance 0;
sid:123;)
```

This is a rule written to detect and alert on a SQL injection attempt to a web server operating on port 80 or 8080. An example would be the following:

```
http://acunetix.php.example/wordpress/wp-content/plugins/demo_vul/
endpoint.php?user=-1+union+select+1,2,3,4,5,6,7,8,9,(SELECT+user_
pass+FROM+wp_users+WHERE+ID=1)
```

The rule starts with the rule action, namely `alert`, which indicates the action that results if this rule matches. The subsequent terms indicate the protocol (`tcp`) that needs to be matched. The rule specifies the TCP destination ports of `80` and `8080`. Typically, these will be HTTP traffic.

The `msg` keyword specifies the message to be included in the generated alert. The `flow` keyword specifies that this rule needs to be applied only to those TCP sessions that are in an `ESTABLISHED` state. Subsequently, the rule goes on to specify that the URI needs to contain certain specific strings.

This gives an idea and example of an IDS/IPS signature. The detailed understanding of such a signature is beyond the scope of this chapter and will be discussed in *Chapter 14*.

Anomaly-based intrusion detection

Anomaly-based intrusion detection detects malicious activity by how it differs from normal behavior. This often requires the system to define and/or learn normal behavior. Since the normal for one environment is often different than the normal for another environment, this approach typically requires a learning phase where the system learns the appropriate normal for a particular environment. During the learning phase, a baseline for normal activity is recorded; subsequently, in the running phase, the activity is compared against the baseline to detect anomalies.

One of the main advantages of this approach is that the anomaly-based approach does not require signatures, and the race against time for security coverage is not an issue. In other words, the anomaly-based approach can detect novel attacks that the IDS/IPS has not encountered before.

On the other hand, the main challenge for anomaly-based systems is that of false positives. Anomaly detection assumes that the outlier case is malicious. However, all outliers are not malicious, and this is the underlying reason for the high false positive rates associated with this approach. Subsequently, significant effort would be required to tune the system – to balance the false positives and false negatives.

Additionally, since the anomaly-based IDS generates alerts when there is a deviation from normal, the alert will not be specific; the system only knows that it is not normal. This results in non-specific or vague alerts being generated.

There are several sub-types of anomaly-based intrusion detection, namely the following:

- **Statistical anomaly-based**: In the statistical anomaly-based approach, the IDS analyzes a set of predetermined values or variables (for example, packet sizes, login session variables, packet header values, and amount of data transferred) and maintains a baseline learned during the learning phase. Subsequently, the system analyzes the set of variables at runtime for deviation from the expected baseline. The system typically has a threshold setting that can be configured, and when the deviation from the predicted baseline is greater than the threshold, it detects the activity as malicious.

- **Machine learning-based**: Machine learning has made significant advances, and this approach is often used to detect outliers. Therefore, the technique is very good for anomaly detection-based IDS/IPS. This is a vast topic, but various techniques under machine learning can be used to detect unknown attacks.

- **Protocol anomaly-based**: This approach applies mainly to network-based IDS. Network traffic typically follows various network protocols. For example, email communication typically follows a set of protocols such as SMTP, IMAP, and POP. These protocols are clearly defined by specifications described in documents called RFC. Protocol anomaly-based IDS detect a deviation of network traffic from the concerned protocol's RFC specification.

Anomaly detection can be a very powerful technique for detecting intrusions since it can detect new and unknown attacks, provided we can overcome the challenges, including high false-positive rates and tuning difficulties. One such technique combines anomaly detection with signature-based detection to create a hybrid solution.

Hybrid intrusion detection

As the name suggests, hybrid IDS combine signature-based and anomaly-based approaches to detect malicious activity. In the simplest design, the network traffic is processed by a signature-based component as well as an anomaly-based component, and the findings of each component are fed into a decision module that makes a final judgment on whether there is an attack or not.

In a more practical sense, typical IDS/IPS will be signature-based but may have some detection modules that work using an anomaly-based approach.

In the next section, let us discuss the state of the art in IDS/IPS. The section will discuss the important features present in the latest IDS/IPS.

The state of the art in IDS/IPS

The intrusion detection and prevention field has been evolving for a few decades. During this period, several commercial and open source IDS/IPS have been developed. As the nature of the internet and its protocols, as well as the complexity of threats, evolved, the IDS/IPS also had to evolve in order to keep up with the threats. Snort is an open source IDS/IPS that was created in 1998, and over the past 20+ years, it has evolved into one of the leading IDS/IPS software. Bro is another open source project, which started in 1994 and was mainly used in an academic setting for several years. Recently, it was renamed Zeek, and a community has formed around the open source project. Suricata is a relatively late player in the game and was created in 2009. It is a signature-based IDS/IPS similar to Snort. The rule syntax for Suricata is very similar to that of Snort. In addition to the rules, Suricata has many other similarities to Snort in functionality – although the design and implementation are completely different.

These three open source IDS/IPS have kept up with the challenges that they faced and stood the test of time. It may be said that the current state of these three IDS/IPS represents the state of the art in IDS/IPS. In this section, let us describe some of the challenges that these systems have faced and what features solved them.

Stateful analysis

Stateful analysis of the various network protocols is a necessary feature in any IDS/IPS. Snort was completely stateless and basically a packet analysis IDS in the initial years. Even when stateful analysis was introduced in the subsequent years, it was incomplete and insufficient. Ideally, the IDS/IPS must analyze the network traffic exactly as the end hosts would analyze it. This means that the IDS would need to maintain a very similar state to the end hosts. This is not a trivial task. This is the reason why it took decades for Snort to improve its stateful analysis functionality. Currently, one could say that Snort is a stateful IDS/IPS device, even though there are still limitations.

Fast packet acquisition

Historically, IDS/IPS devices used the packet capture library called libpcap. This is a library used by the tcpdump project and was available as open source. libpcap worked great, but as the internet speed increased, this library started becoming a performance bottleneck. In the case of libpcap, the packet data (network traffic) had to be copied several times before reaching the IDS for processing, and this was one of the reasons for the performance issue. Currently, the state of the art uses zero-copy mechanisms in order to improve performance. Although Snort still supports and offers libpcap-based packet acquisition, it offers all the latest packet acquisition mechanisms to be used.

Parallel processing

The state of the art is for the IDS/IPS to perform network traffic analysis using parallel processing – this could be a multi-process-based or multi-thread-based design. Snort started as a single-threaded, single-process IDS, and then evolved into a multi-process design. Currently, Snort uses a multi-threaded design.

In a multi-process and multi-threaded design, an incoming session would be processed by one of the processes or threads. Once a session is analyzed by a process or thread, then all the subsequent network packets for that session will be analyzed by that process or thread. This is called session pinning. Typically, such pinning is based on a hashing approach, where the hash will be based on the source and destination IP addresses, port, and protocol. However, in this approach, two related sessions that hash to two separate processes or threads will result in a lesser-grade analysis.

Pattern matching

Pattern matching has been and is still one of the most important features of IDS/IPS. A single signature may contain several pattern matches. Originally, these were evaluated one rule at a time, one pattern at a time. With time, multi-pattern search algorithms were used in order to speed up the rule processing.

In addition, as opposed to the crude pattern matching of the past, current IDS/IPS devices perform the pattern matching with context. For example, when a pattern is specified, it can also be specified what data to match against – HTTP URI, HTTP header, and so on. This improves the performance since the pattern search can be limited to specific data, and it also improves detection accuracy.

Extending rule language

Most IDS/IPS have a rich rule language. However, there will always be cases that cannot be covered by the limited capability offered by the rule language. Each system – Snort, Suricata, and Zeek – has its own approach to this challenge. Zeek from the Bro days had a full-fledged language to write detections in. So, the challenge really did not apply to Zeek. Snort came up with **shared object (SO)** rules, whereby custom C code could be written for a particular functionality and released as .so files in a release. Suricata integrated Lua scripting as part of the rule language extension.

App and protocol identification

Historically, Snort rules were based on protocol and port. For example, the rule would specify that it applies to TCP and on ports 80, 8080, and 3128. The list of ports could be more extensive to cover the usual HTTP ports. However, if there is an HTTP session on port 1000, the rule will not be applied against that session. This challenge was solved by introducing the app and protocol identification feature, which is a state-of-the-art feature. All leading IDS/IPS detect the various protocols on any random port to perform the analysis correctly.

File analysis

In certain cases, the IDS/IPS must analyze the data not as a stream of bytes but as a file. This feature is also the state of the art and is part of all leading IDS/IPS.

These standard features represent the state of the art in IDS/IPS. The intent of this discussion was not to present a comprehensive set of features but to give an idea of the various features.

The next section discusses the various metrics that are used to evaluate IDS/IPS. These metrics try to measure how effective the IDS/IPS are from an accuracy perspective, as well as how efficiently they do their tasks.

IDS/IPS metrics

It is essential to be familiar with a few key metrics that are often used to describe how capable an IDS/IPS is. An IDS/IPS has two main metric classes: detection accuracy and performance metrics. These metrics are mainly used to compare IDS/IPS, which are also known as IDS/IPS evaluations. We will look at these topics in this section.

Detection accuracy

Every packet or connection analyzed has two possibilities – benign or malicious. Also, there are two possibilities for IDS analysis results – an alert is generated or no alert is generated. So, we end up with four possibilities, as described in the following table. This table is called a confusion matrix and is a valuable way to measure the performance of an IDS in classifying the connections or sessions as benign or malicious.

	Benign	Attack
No alert generated	True negative	False negative
Alert generated	False positive	True positive

Table 1.1 – Intrusion detection confusion matrix

Let's look at each of these cases:

- **True positive (TP)**: TP is the case when the connection is malicious and the IDS correctly alerts.

- **True negative (TN)**: TN is the case when the connection is benign and the IDS correctly avoids generating an alert. Ideally, the TN rate should be 100%; this means that 100% of benign connections will result in an absence of an alert.

- **False positive (FP)**: FP is the case when the connection is benign and the IDS incorrectly generates an alert. Ideally, the FP rate should be 0%, meaning that the IDS does not generate any alerts for benign connections.

- **False negative (FN)**: FN is the case when the connection is malicious and the IDS incorrectly fails to generate an alert. Ideally, the FN rate should be 0%, meaning that the IDS does not fail to generate an alert for malicious connections.

Now, the metrics used for detection accuracy are as follows:

- **True positive rate (TPR)**: The TPR is calculated as the ratio of accurately detected attacks (TP) to the total number of attacks (TP + FN). Note that the total number of attacks is equal to the number of attacks detected (TP) plus the number of attacks missed (FN).

$$TPR \ = \ \frac{TP}{TP + FN}$$

  Ideally, the TPR should be equal to 1, which means FN should be 0; in other words, the IDS alerts on all the attacks.

- **False positive rate (FPR)**: The FPR is calculated as the ratio of wrongly detected attacks (FP) to the total number of benign connections (FP + TN).

$$FPR \ = \ \frac{FP}{FP + TN}$$

  Ideally, the FPR should be equal to 0, which means FP should be 0; in other words, the IDS does not alert on any of the benign connections.

- **Precision rate (PR)**: The PR is calculated as the ratio of accurately detected attacks (TP) to the total number of alerts generated (TP + FP).

$$PR \ = \ \frac{TP}{TP + FP}$$

  Ideally, the PR should be equal to 1, which means TP should be non-zero and FP should be 0; in other words, all the alerts generated by the IDS should be for attacks.

The preceding metrics are useful in measuring the detection accuracy of IDS or IPS. The values of these metrics are useful for comparison purposes (to choose one system over another) or for benchmarking purposes (to measure the improvement of a system over time).

Generic versus specific signatures – a discussion

For a signature-based IDS/IPS, these metrics mostly depend on how specific or generic the signatures are. If a signature is too generic, it tends to have a good TPR, but at the same time, the FPR also increases. On the other hand, if a signature is too specific, it will result in low FPRs. However, it will also result in a miss when there is a slight modification to the attack; that is, FN increases.

Performance-related IDS/IPS metrics

The traffic volume on the internet, as well as the traffic volume within any network, has been increasing year by year. *Figure 1.6* shows how the traffic volume has increased across the internet over the past several decades.

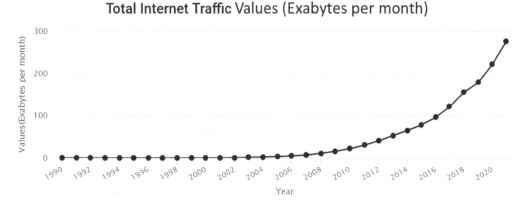

Figure 1.6 – The traffic volume on the internet

In addition, the network IDS/IPS is typically deployed in key points in the network where it must monitor the traffic to and from the entire network. IDS and IPS must perform at an efficient pace to keep up with the increasing traffic loads.

The complexity of the analysis performed by the IDS affects this rating. More complex analysis (for example, a higher number of signatures to check, or more complex signatures to check) leads to an increase in the IDS processing time for a packet. Typical IDS and IPS have configurations, which include various parameter settings that control their behavior, as well as the total database of signatures to match against. By controlling the configuration, we can control the packet processing time of the IDS/IPS, thereby affecting the throughput that can be sustained.

The following metrics are often used to measure the performance of an IDS/IPS:

- **Throughput**: This is the maximum amount of network data that can be analyzed by the IDS without packet drops. This is measured in bits per second (or megabits per second or gigabits per second).

- **Latency**: This metric is only applicable to IPS devices since it works in an inline (not offline, passive) fashion. The network traffic traverses the IPS, and packets are forwarded only after the IPS has evaluated it. This introduces a delay in the network traffic, which is measured by this metric. The higher the latency, the worse the performance of the system. This is typically measured in nanoseconds or microseconds.

- **Packets per second**: This is the maximum number of packets per second that can be analyzed by the IDS without packet drops. This is measured in the number of packets per second. Not all packets are the same; some packets take more time to be analyzed than others. So, this number has to be measured while maintaining the traffic profile as normal as the IDS would typically analyze.

- **Packet drop rate**: This is the rate that indicates the number of packets that are dropped by the IDS. This is usually specified in the number of packets per second.

- **TCP connections per second**: This is the rate of TCP connections that can be analyzed by the IDS. This is measured in connections per second.

- **Simultaneous TCP connections**: This metric indicates the number of TCP connections that the IDS can analyze simultaneously. To analyze a TCP connection, the IDS needs to maintain the TCP state and other data structures, which consume memory. Subsequently, this metric indirectly measures how much memory capacity the IDS has.

The preceding IDS/IPS metrics are useful for the performance evaluation of the system. In order to enable businesses to operate well as well as to provide protection, IDS and IPS devices must be highly efficient.

IDS/IPS evaluation and comparison

IDS/IPS evaluation is a process that involves a series of tests and/or experiments in order to measure the detection accuracy as well as the performance of the system. DARPA evaluations concentrated on detection accuracy during the early years of IDS evolution. Organizations such as ICSA Labs and NSS Labs conduct a series of tests that measure the detection accuracy as well as performance ratings of IDS.

These evaluations have to be taken with a grain of salt since the results will depend very much on the selection of attacks as well as the selection of traffic profiles. However, these results are still highly beneficial and help companies narrow down the IDS/IPS solutions to be evaluated in their environment with their particular traffic and test conditions.

Next, let us look at one of the challenges faced by IDS/IPS – IDS evasions. This is the scenario where the attacker is able to conduct an attack through the IDS without getting detected.

Evasions and attacks

An advanced adversary may attack the monitoring infrastructure itself so that their actions are not detected. These approaches may involve the use of evasive techniques to trick the IDS/IPS and avoid detection. Alternatively, they may target the IDS and IPS device itself to render them less effective and thereby avoid detection.

IDS/IPS evasions

IDS/IPS evasion is a technique used by an adversary to trick the IDS or IPS to conclude that there is no attack occurring when there in fact is (evasion), or to conclude that there is an attack occurring when there in fact isn't (insertion).

IDS and IPS are separate entities from end hosts, and there are inherent differences in what network traffic they see and how they process the traffic. Due to these differences, it is possible to craft the traffic in such a way as to trick the IDS or IPS device.

An example of an evasion case is as follows:

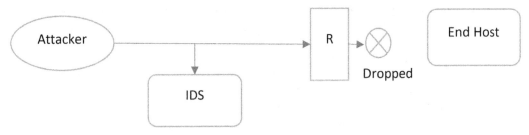

Figure 1.7 – An IDS evasion example

Figure 1.7 shows a typical scenario for an IDS evasion. The box marked **R** is an internet protocol router. According to the IP protocol, as the IP datagram is being routed from the source to the destination, the IP datagram's **time-to-live** (**TTL**) value is decremented by 1 at each router.

In this case, the attacker manipulates the packet TTL values such that all of them are seen by the IDS, but only some of them reach the end host. Thereafter, the attacker sends the following:

1. Packet 1: Data: "ATT".
2. Packet 2: Data: "END" (TTL 1).
3. Packet 2 (Retransmitted): Data: "ACK" (TTL 10).

The attacker sends the attack in two separate packets. Let us imagine that the IDS favors older data when reassembling the segments. So, the IDS reassembles the data (concatenates Packet 1 and Packet 2) as *ATTEND*. On the other hand, due to the TTL manipulation, the second packet does not reach the end host. So, the end host reassembles the data as *ATTACK*, which means a successful attack. The IDS, however, fails to generate an alert because it concluded the data as *ATTEND*.

If the attacker knows that the IDS does not validate packet checksums, the same test as the previous can be repeated, and instead of manipulating the TTL values, they can send the second packet with an invalid checksum, which will be processed by the IDS but discarded by the end host.

The technique for evasion can be adjusted based on the type of difference between the IDS processing and endpoint processing.

Attacks against the IDS/IPS

We saw that the IDS and IPS play a key role in the security posture of an organization. We noted earlier that IDS and IPS processing is complex and involved. In the previous section, we discussed how the adversary tries to go unnoticed during an attack by evasion (and insertion) techniques. In another tactic, the adversary could attack the IDS/IPS itself to render it partially or completely useless for a short or long duration. Such attacks against the IDS/IPS can be classified into two main types:

- **Crash attack**: This is a class of attack that tries to send some network traffic that causes the IDS to crash when processing it. For example, the IDS could have a buffer overflow vulnerability in its decoder or detector module, which is triggered when processing a certain type of traffic. A very early version of Snort (version 1.8) has a similar vulnerability when processing RPC traffic, causing a buffer overflow. An attacker knowing such a vulnerability could target the IDS and cause a crash, which would lead to degraded operation and missed attacks during that time window.

- **Denial-of-service attack**: In this approach, the attacker sends traffic that causes the IDS to spend a large amount of time processing it. A lot of such traffic can cause the IDS to go to a state where it cannot keep up and will start dropping packets. This can lead to degraded operations and missed attacks.

A robust IDS design would consider such attacks against itself and have mechanisms to defend against such attacks.

Summary

This chapter provided a brief introduction to IDS and IPS, including the need for them and the role these systems play in the defense-in-depth strategy. The chapter then discussed the different types of IDS, the current state of the art in the field, and some of the key metrics used to evaluate IDS and IPS.

In the following chapter, we will look at Snort, one of the most popular open source IDS/IPS and discuss the evolution of Snort from its early stage to where it is now.

2
The History and Evolution of Snort

Snort has come a long way – from its humble beginnings in the late 1990s to its current position as one of the most popular open source software of all time. Snort was selected into InfoWorld's Open Source Hall of Fame in 2009. Since then, it has continued to grow and evolve as an open source system and as a network intrusion detection and prevention system. In this chapter, we will discuss how Snort originated and how it evolved into its current form. We will do this by discussing a few of the key releases of Snort and their features.

The main topics covered in this chapter are as follows:

- The beginning of Snort
- Snort 1 – key features and limitations
- Snort 2 – key features, improvements, and limitations
- The need for Snort 3

The beginning of Snort

Intrusion Detection Expert System (IDES) was one of the first IDS developed at SRI International in the late 1980s. By the 1990s, many innovative IDS were being used, including **Network Anomaly Detection and Intrusion Reporter (NADIR)** and **Network Security Monitor (NSM)**. **Network Flight Recorder (NFR)** was one of the early systems that worked using the libpcap packet capture library, and NFR provided stateful packet inspection, misuse detection, and protocol anomaly detection. NFR was, however, a commercial system and was not available for use to the public. The Lawrence Berkeley National Laboratory released Bro in 1998 – a network IDS that also used libpcap; Bro specified policies and rules using a custom rule language.

In those days, around 1998, Martin Roesch created a program that filled an important *ecological niche*, in his own words, which he initially named APE and then renamed Snort. The author released Snort to the public under the **General Public License (GNU)**. In the beginning, Snort had humble goals: it was meant to be a *"tool for small, lightly utilized networks"* and to be used *"when it is not cost efficient to deploy commercial systems."*

Snort was made available to the public via Packet Storm (www.packetstormsecurity.com) and one of the most attractive features of Snort was the ease with which users could add and contribute to the collective Snort rules. A community called **Snort Community** was formed very quickly; by 2008, it had 300,000 active users.

Snort rules were one-dimensional and simple to write. This worked very well with users who had to add rules for their own custom environments or to add rules quickly as a response to an immediate security threat.

Let us now delve into some of the Snort releases of the past, starting with Snort release 1.0, which came out in 1998, and discuss the relevant features that were available as part of the release. This will give an idea as to how Snort evolved with time and with the evolution of the security threat landscape.

Snort 1 – key features and limitations

Snort 1.0 was released in April 1999. It was the year of the Melissa email virus. The ping of death, Smurf, **Local Area Network Denial (LAND)** attacks, and website defacements were some of the threats of the time. Detecting such attacks only required basic decoding of the various packet headers.

Snort 1.0 had very limited features. The main features of Snort 1.0 included packet decoding functionality, a detection engine (for matching packets against rules), and the feature to create alerts/logs. The Snort rules capability was also limited and supported on the following keywords: content, msg, flags, ttl, itype, and icode.

The Snort code base contained just 10 files and 5,000 lines of code: A sample rules file was part of the Snort 1.0 release and had 18 signatures, and a subset of that is shown as follows. Notice the simplicity of the signatures:

```
alert tcp any any -> 192.168.1.0/24 any (msg:"SYN-FIN scan!"; flags:
SF;)
alert tcp any any -> 192.168.1.0/24 any (msg:"Null scan!"; flags: 0;)
alert tcp any any -> 192.168.1.0/24 143 (msg:"IMAP Buffer overflow!";
content:"|90E8 C0FF FFFF|/bin/sh";)
```

That's it – Snort 1.0 was as simple as that. The simple and easy-to-understand system combined with the ease with which users could add and extend their signature database made it very attractive to security administrators and users.

Snort 1.5 was released around December 1999, and this release provided the basic architecture of Snort that became the foundation of Snort for quite some time. Snort 1.5 introduced the plugin architecture and added a couple of basic preprocessors such as the `minfrag` preprocessor (a module that detects small IP fragments), and the HTTP decode preprocessor (a module that decoded percent-encoding in HTTP requests). This release aimed at detecting some of the notable threats of the time including portscans, SYN flooding, IP spoofing, TCP sequence number attacks, TCP session hijacking, RST and FIN attacks, Ping of death, and vulnerabilities of the common protocols over TCP/IP (namely, SMTP, Telnet, NTP, Finger, and WWW).

As an IDS, Snort was still catching up on features necessary to operate and detect the latest trends in cyber-attacks. Notably, Snort was still missing TCP stream reassembly modules necessary to detect attacks in the presence of tools such as fragrouter/fragroute, which split the attack data across several TCP segments and cause packet-based detection to miss the attack.

Snort 1.7 was released in January 2001. It was in this release that the first versions of the TCP state tracking and stream reassembly modules were included. Another key component that was included in this release was the IP `defrag` module, which assembles the IP fragments into a single datagram before running through the detection engine. The release also had an improved HTTP preprocessor. Snort 1.7 added some key features such as TCP stream reassembly and IP defragmentation modules to the Snort code base. These modules continued to evolve and were revised as Snort continued to evolve. In Snort 2.8, both the TCP stream reassembly and IP fragmentation modules were improved.

Snort 1.8 was a key release in the evolution of Snort. This release had several minor version releases, from Snort 1.8.1 to Snort 1.8.7, and these were released over a year (Snort 1.8 was released in July 2001 and Snort 1.8.7 was released in July 2002). The main changes in Snort 1.8 included updates to the TCP state tracking and stream reassembly preprocessor, the IP defragmentation preprocessor, and the introduction of Telnet and RPC preprocessors.

The only significant feature to note as part of the Snort 1.9 release is the *flow* feature. This, however, is a key development and one that most of the Snort users still recognize. This feature added a new rule option called `flow`, which is specified in all of the TCP Snort rules. This feature added to the *stateful* nature of Snort in those days. If a Snort TCP rule stated `flow:established`, then that rule would only trigger on TCP sessions that were in the `TCP ESTABLISHED` state. In addition, `flow` could specify the direction of the flow; for example, a rule could specify `flow:established,to_client`, which meant that this rule would only trigger on ESTABLISHED TCP sessions and only for traffic from server to client.

Till now, we have looked at various releases in the Snort 1.x series. These releases happened within a span of approximately 3 years. Snort 1.9 was released around March 2003. By this time, Snort also had evolved as a network IDS and added key features and functionality. A significant user base was also formed resulting in thousands of downloads. By 2003, information security had become a business problem from a purely technical problem. The number of security issues and incidents was on the rise, and there was a need for additional features from an IDS point of view. In addition, there was an increasing need for a high-speed IDS. Snort 1.x was more about adding features and not about high performance. Snort 2.x releases focused on these needs.

Let us continue to discuss the later released Snort versions from 2.0 onward.

Snort 2 – key features, improvements, and limitations

In this section, we will continue to study how Snort continued its evolution. Snort 2.x releases continued to add key features to Snort functionality. However, there was significant importance given to *performance* – the in-depth analysis had to be done in an efficient way so that the system could keep up with increasing network traffic loads.

The number of signatures was going up, and the performance of the IDS was affected. Snort 2.0 came with a significant enhancement – introducing multi-pattern search. Multi-pattern searches such as the Aho-Corasick algorithm enabled an $O(1)$ search across all the signatures for a given packet or stream data, and this resulted in a subset of signatures that needed to be completely evaluated. This gave a significant performance improvement for Snort. In addition, Snort 2.0 introduced the concept of HTTP flow-based analysis.

The main features of Snort 2.0 were the addition of an enhanced detection engine designed for high performance, a new HTTP analyzer preprocessor, enhanced protocol decoding and protocol anomaly detection, and an alert thresholding/suppression capability.

Snort 2.1 came with a few powerful features as well, including a new HTTP preprocessor called HTTP Inspect. The first version of the HTTP preprocessor was very specific to performing a set of encoding such as percent encoding. With the HTTP Inspect preprocessor, Snort's analysis of the HTTP protocol improved significantly. The Snort 2.1 release also added the ability to use regular expression matching in signatures (using `pcre`). In addition, this release introduced the `flow` keyword. This feature detects and classifies the TCP communication as two flows, namely, client to server and server to client. Subsequently, signatures could be written to match a particular flow. In addition, the `flow` keyword can also be used to match the state of a flow; it leverages the state maintained by the TCP stream reassembly module for this purpose.

Snort 2.2 did not contain many significant features, and we will not discuss the respective changes. In the next section, we will discuss Snort 2.3, which added a key feature. Till now, Snort was only an IDS, but with the next release, Snort was evolving into an IDS/IPS.

Snort Inline was a project initiated by Will Metcalf and Victor Julien. The project patched the Snort source code to convert it into an IPS. They started with Snort 2.0.6 and made the patch available with each Snort version. These changes from the Snort Inline project were incorporated into mainline Snort with the release of Snort 2.3. This was a significant milestone – Snort made the leap from being an IDS to being an IDS/IPS.

In addition to the IPS functionality, the other major changes in release 2.3 included a new portscan detector module and a mini-preprocessor to catch the X-Link2State vulnerability.

The Snort 2.x code base seemed to become more stable. The releases contained more bug fixes and optimizations than new features when compared to initial versions.

Snort release 2.4 contained performance improvement changes for the flow and stream4 preprocessor session management in order to limit memory usage. In addition, the IP fragmentation reassembly preprocessor was upgraded (Frag3 preprocessor). A new detection plugin was also added to detect FTP bounce attacks.

One of the challenges of IP fragmentation reassembly modules is that in some of the use cases (for example, overlapping IP fragments), the IDS had to analyze a certain way; however, the end host may be doing the analysis in a different way, and this leads to evasion possibilities. In case of IP fragment overlap, during the IP fragmentation reassembly, the IDS must favor either the original bytes (first received by IDS) or the retransmitted bytes (received later). If the IDS *always* favored the original or the retransmission for the reassembly operation, then its analysis could be different from how the endpoint does it. This, as mentioned earlier, can lead to evasion scenarios. With the introduction of target-based behavior, the IDS could be configured with information as to which end host (IP address) is which OS and which behavior is favored. This would lead to an analysis that was consistent with the end host, thus lowering the chances of an evasion.

Snort 2.6 was released with a bunch of features including some core architectural changes. This included the performance profiling feature and the dynamic plugin architecture. The release also supported target-based TCP stream reassembly and IP fragmentation reassembly functionality.

With Snort 2.8, the project seemed to have more dot releases; for example, Snort 2.8 had up to 2.8.6, and Snort 2.9 has had up to 2.9.18 (and more to come, maybe). As before, we will combine the features that were released in all dotted releases as Snort 2.8 and Snort 2.9.

Snort 2.8 had several important features, including IPv6 support and HTTP detection. Snort 2.8 also introduced the `fast_pattern` keyword to specify which content to use as the MPSE fast pattern. In addition, the release added the capability to reload IDS/IPS configuration without a restart, and the ability to support multiple configurations.

We will discuss the Snort 2.9 release as a subsection as this was the final release of the Snort 2.x series and the main release before Snort 3.

Snort 2.9

Snort 2.9 is the final Snort 2.x release. It has had 18 dotted releases; the current and latest release is 2.9.18. Let us briefly look at the features that Snort 2.9 introduced:

- **Data Acquisition (DAQ) layer/library**: Until now, the packet acquisition was not abstracted out yet. The various libraries used for getting packets from the network (via or without kernel involvement) were tightly integrated into Snort. With this feature, the packet acquisition happened via a single and consistent API. The various libraries and their details were abstracted out from the Snort code and moved into the DAQ library. This was a welcome change to make the system more modular, as well as to make the Snort code and system simpler.

- **TCP state tracking and stream reassembly – inline operation**: The way TCP stream reassembly and state tracking must be done is different when Snort is deployed in an IDS versus IPS way. With this release, Snort optimized this module for IPS deployments. A new preprocessor was also added, which is relevant when Snort is deployed as an IPS; this module handled packet normalization to allow Snort to interpret a packet the same way as the receiving host. This was a significant step toward tackling evasion attacks.

- **Rule engine enhancements**: The detection capabilities continued to increase with the addition of `byte_extract` and `byte_math` features, as well as Base64 decoding ability.

- **Protocol Aware Flushing (PAF)**: This feature enhanced the TCP stream reassembly in a significant way and made the detection more accurate. For example, in the case of HTTP, if the connection is kept alive, the TCP stream reassembly operation should clear out all the data of prior HTTP requests before reassembling the next HTTP request and sending it for matching against signatures.

- **New preprocessors**: Preprocessors are the heart of protocol decoding and the enabler for signature matching. It is the preprocessors that do the headwork so that the rule matching can happen.

    With Snort 2.9, a set of new preprocessors was added, namely the following:

    - SIP preprocessor to analyze SIP-related traffic.

    - POP3 and IMAP preprocessors to decode and analyze the mail protocols, IMAP and POP.

    - IP reputation preprocessor enabled Snort to blacklist or whitelist specified IP addresses.

    - SCADA (DNP3 and Modbus) preprocessors.

    - GTP decoding.

Snort 1.x releases spanned a period of 3 years. Snort 2.x releases covered a period of 19 years (and counting). At the end of Snort 1.x releases, Snort was still maturing as an IDS. At the end of Snort 2.x releases, Snort is a matured IDS/IPS system and one of the best of its class in the market. Snort 2.x still has room for improvement.

One of the key challenges of Snort 2.x is the fact that, at the core, it is a single-threaded analysis engine. If one needs to perform high-speed intrusion detection using Snort 2.x, the solution involves running multiple instances of the Snort process. This can be done, although it has certain disadvantages from a security analysis and detection point of view.

Over the past years, from Snort 1.0 to Snort 2.9, the system has grown more and more complex. In order to take Snort to the next level and be the best in the field, there was a need to make a few architectural changes such as making the system more modular and making the code, configuration, and overall system simpler.

The need for Snort 3

As we briefly saw in this chapter, Snort started with a very humble goal. As the networks, protocols, and the nature of security threats evolved, Snort also grew and evolved. It is not the IDS for small networks anymore! Snort can stand against any commercial or open source IDS/IPS as of now. It is one of the best in the IDS/IPS space.

That said, the nature of threats and the internet itself continues to change, and Snort has to evolve as well. Snort 3 has been cooking for a long time. Snort versions 1 and 2 have been single-threaded. For higher performance, the solution was to run multiple instances of Snort. However, this has several challenges, which are as follows:

- The multiple instances of Snort do not share state between them. This limits the possibility of information sharing between sessions and improving detection.
- Performance challenges.
- Various limitations specific to individual modules, such as the following:
  - Complex implementation of modules such as TCP stream reassembly.
  - Partially stateful HTTP module.

This should help give you a brief idea of the need for Snort 3.

Summary

In this chapter, we took a journey through the evolution of Snort. It gave you a high-level idea of how features were added and modified as the years went by. It gave a glimpse of how Snort 1.0, which started off with a small goal, became Snort 2.9, one of the best IDS/IPS in the field currently. We have not listed each and every feature that was part of Snort releases and have combined the features of dotted releases together with the goal of simplifying the journey. In the next chapter, we will investigate and learn the details of Snort 3.

Part 2:
Snort 3 – The New Horizon

The second part of the book introduces you to Snort 3, which is a significant milestone in the evolution of Snort. In this part, we will discuss the system architecture and main components of Snort 3. You are also guided through the step-by-step process of Snort installation. These steps will be discussed and elaborated on, with two operating systems as examples. You will also be introduced to the Snort 3 configuration in the final chapter of this section.

This part has the following chapters:

- *Chapter 3, Snort 3 – System Architecture and Functionality*
- *Chapter 4, Installing Snort 3*
- *Chapter 5, Configuring Snort 3*

3
Snort 3 – System Architecture and Functionality

Snort 3 is a significant milestone in the evolution of the Snort IDS/IPS project. Snort 3 was under development for a long time and has finally come to reality and general availability. In the last chapter, we discussed the evolution of Snort from inception till now – from version 1.0 to version 2.9. Compared to Snort 2.0, the number of changes introduced in Snort 3.0 is significantly higher; in other words, Snort 3.0 is a giant evolutionary leap in Snort's growth. Snort 3.0 introduces changes to rule syntax and language that are not compatible with previous Snort versions; it introduces Lua-based configuration that is not compatible with Snort 2.x. In addition, there are key architectural changes to the system to make it highly modular and push the limits of a high-speed IDS/IPS. In this chapter, we will discuss the following topics:

- Design goals
- Key components
- Snort 3 system architecture

First, let us look at the design goals for the Snort 3 project.

Design goals

The concept and idea of what became Snort 3 was started in 2005 under the name **SnortSP** (short for **Snort Security Platform**). The name suggests that the authors were trying to build a platform for the next-generation IDS/IPS. Later, this project was internally dubbed Snort++. As this project progressed, many of its features were pulled into the mainline Snort releases. However, the architectural changes could not be incorporated into Snort 2.x as these were foundational changes. These are the changes that have been packaged as Snort 3, which we will discuss in this section.

The main design goals of Snort 3 were as follows:

- High performance.
- Modular and pluggable architecture.

- Better configurability.
- Efficiency.

High performance

Snort has historically been a single-threaded monolithic program. This meant that, at any time, one Snort program would process and analyze only one packet (or one stream) at a time. This became a challenge as the network throughputs of typical networks kept increasing, and Snort had to keep up with the traffic in both passive mode (IDS) and inline mode (IPS). In order to scale up and process much higher network throughputs, the straightforward solution was to build a system with multiple CPU cores and run multiple Snort instances to process several packets (or streams) simultaneously. In order to distribute network traffic to the various Snort instances, DAQ modules such as PF_RING came in handy.

Alternatively, there was the option of using a multithreaded architecture, which had several benefits, which we will discuss in the following subsections.

Better memory usage

One of the challenges of running multiple instances of Snort was the resultant high memory usage. Each instance of Snort would parse the same configuration and signature set and create the necessary data structures in memory. If the system has 16 CPU cores and we run 16 Snort instances, the memory used to store the rule engine data structures and similar configuration data is duplicated 15 times!

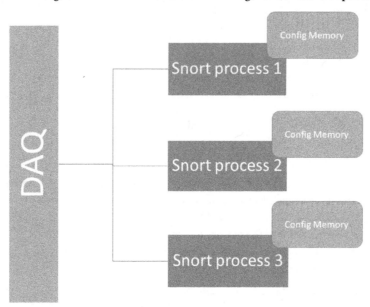

Figure 3.1 – A system running three instances of Snort

Figure 3.1 shows a system running three instances of Snort. As each of the Snort processes has a copy of the memory used for storing configuration data including Snort signatures, there is duplication of memory.

Implementing a multithreaded system would solve this issue. The saved memory could be used for other purposes. This system architecture is therefore employed by Snort 3, as we shall see in the *Snort 3 system architecture* section.

State sharing

In the multi-process architecture, which is used in conjunction with Snort 2, each Snort instance is a separate process. The network traffic is intelligently divided among these instances using various strategies. One of the common strategies for splitting network traffic is based on the five-tuple hash – namely, source IP address and port, destination IP address and port, and IP protocol (TCP/UDP/ICMP). This approach works well when the traffic sessions are not related to each other. However, in reality, the traffic includes network flows and/or sessions that are related to each other, and these may be split and distributed for analysis to separate Snort instances. This results in an incomplete analysis; hence, there is a chance that the system might miss out on certain attack scenarios.

A simple example that explains this point is that of a TCP port scan where the attacker sends TCP connection requests to determine which TCP ports are open. To simplify, let us say the attacker only scans a single host and scans 16 well-known ports (for example, 20, 21, 22, 23, 25, 53, 79, 80, 110, 119, 143, 135, 138, 139, 443, and 445):

- In the case of a single Snort instance processing these TCP connection requests, the port scan detection module sees all 16 TCP connection requests. Based on the configuration settings, it may detect a port scan.
- In the case of a system with 16 CPU cores running 16 Snort processes and with a DAQ that splits the traffic among these processes by hashing the 5-tuple, each Snort process will see only a small subset of the whole scan. In the perfect traffic-splitting case, each Snort process will see only one TCP connection request.

The multiple threads approach enables state sharing between the different Snort processing threads – which is a great win. Port scan detection is just one of the many cases where state sharing is necessary for better detection. A few of the features that exist in the current state-of-the-art IDS/IPS (Snort, Suricata, and so on) that are affected by this architectural decision are alert thresholding (detection filters) and global flowbits. Similar to the port scan detection problem, alert thresholding also requires rate-based calculation. If the alerts are generated by separate Snort processes that do not share data, accurate alert thresholding is not possible. However, in Snort 3, the multiple Snort threads can share data and state, thereby enabling various features that require such state sharing.

Pluggable modular architecture

The Snort 2.x program consists of various modules, such as decoder modules, preprocessor modules, detection plugin modules, output plugin modules, and so on. In earlier versions of Snort, all these modules were statically compiled to create the Snort program. Any addition to the functionality meant that a new Snort would have to be compiled. For example, when there is a change to the TCP state tracking (stream) preprocessor, it required a new build of Snort.

The Snort team may have envisioned a pluggable modular architecture for the future. As Snort evolved, as part of the Snort 2.6 release, the concept of dynamic modules (dynamic preprocessors, dynamic plugins, and so on) was introduced. Snort 2.6 was released in 2006, while the SnortSP project started in 2005. It is possible that the idea may have been shared between the projects.

The idea of pluggable modular architecture makes the system more adaptable. It will be very easy to replace any of the modules with third-party modules, and the approach encourages more community participation in the project's development.

Event-driven approach

In Snort 2.x, the preprocessor modules used callback functions for their operation. For example, each preprocessor would register a callback function for setup, configuration, packet processing, and so on. When Snort initialized, the setup callback function of each preprocessor was invoked. During configuration time, the configuration callback function was invoked. During the packet processing time, the packet processing callback function was called for every incoming packet or stream. If a packet was not meant for a preprocessor, it would just exit after a check. The preprocessors would do the analysis irrespective of whether the result was used by subsequent modules or not. This is called the **just-in-case** approach.

Alternatively, Snort 3 uses an event-driven publisher-subscriber approach. In this approach, the module does its work only if there are subscribers. If there are no subscriptions, there is no need to do the analysis. This is called the **just-in-time** approach (as opposed to the just-in-case approach in Snort 2.x).

Configurability and customizability

The Snort configuration file syntax and the Snort rules syntax have been consistent from the initial versions till the Snort 2.9 release. The configuration file and rules syntax were sufficient for the early Snort, which began as an IDS for small networks; since then, Snort has developed into the current state-of-the-art IDS/IPS. Subsequently, there was a need to update the capability of the system from a configurability standpoint. The goal was to build a system that would enable the user to configure it easily.

The goal was to make the rules and configuration more readable, meaningful, and contextual. In addition, there was a need for it to be scriptable. The answer to these can be found in LuaJIT.

Efficiency

Snort has come a long way, growing from a lightweight IDS for small networks to one of the best IDS/IPS in the field. This was accomplished by adding features and functionality to the system over several years and over several releases.

This usually leads to the system becoming massive and slow in processing speed. The Snort architecture team has occasionally made changes to the system to address this aspect. For example, the changes to the detection system, multi-pattern search, and so on were introduced to increase efficiency.

The Snort3 project had been brewing for quite some time. Many code edits or changes to Snort 2.x, that were related to efficiency, were borrowed from the Snort 3 project. Efficiency was, in fact, one of the key goals of the Snort 3 project – to make some core changes to the system so as to make the entire system efficient so that the Snort IDS could do what it does faster, consuming fewer resources. In some ways, the *improved efficiency* goal is also related to the *high-performance* goal of Snort.

In the next section, we will look at the key components of Snort 3 and we will discuss how they satisfy the design goals we discussed in this section.

Key components

Snort 3 has gone through an extended incubation period. Based on the Snort 3 project's design goals, the system was built and finally released to be generally available in January 2021. The main components of Snort 3 are as follows:

- Packet acquisition module (DAQ).
- Codecs (decoding) module.
- Inspectors (preprocessors).
- Detection module (rules engine, **Shared Object (SO)** rules).
- Alert and logging module.
- Configuration module.

Let us look closely at each of these modules, starting with the **Data Acquisition (DAQ)** module.

DAQ module

The DAQ module, also known as the **Data Acquisition (DAQ)** module, is responsible for acquiring the network traffic and providing it to the rest of Snort for analysis. It was first released as part of the Snort 2.9 release. The packet acquisition functionality was fulfilled by the `libpcap` library for a long time and was later abstracted out into the DAQ layer.

The DAQ module boasts a versatile library (LibDAQ) that exports various API functions. Snort uses these functions without needing to delve into the intricacies of the underlying mechanisms. Irrespective of whether the network traffic is being read from a **packet capture** (**pcap**) file or from the IP Tables Firewall interface (inline mode), Snort can employ the same API to initialize and read the network traffic.

The DAQ abstraction layer also enables third parties to create and make available DAQ modules for custom solutions. Third-party DAQ modules include the PF_RING and Napatech modules, among others.

Codecs

The DAQ module presents the raw packets as captured from the various interfaces. This data is in the form of raw bytes that have not been decoded or interpreted.

This layer was traditionally known as the decoder layer. Its job is to start decoding the bytes starting from the lowermost layer. For example, if the **IP QUEUE** (**IPQ**) module is selected as the DAQ module, then the received bytes will be interpreted as an IP datagram, and so on.

In Snort 3, the various Codecs are entirely pluggable. This enables anyone to rewrite any available Codecs and use them as a pluggable module if they deem it appropriate.

The name *codecs* come from the terms *code* (encode) and *decode*. The decode function interprets the raw bytes according to the relevant protocol specifications. The encode functionality creates the bytes to send out to the network. This happens when the system in IPS mode sends out TCP RESET packets to terminate a connection or modifies a packet by rewriting the content. We will learn more about these in a later chapter.

Inspectors

This module replaces the traditional preprocessor module. The inspector's essential role is similar to that of a preprocessor: to perform protocol-specific analysis, normalizations, and state tracking so as to enable the detection module to perform rule matching accurately.

The Snort 3 inspectors use the publisher-subscriber mechanism to do their tasks and publish the output for any module that may be interested in the processed data. This enables just-in-time rather than the just-in-case processing used in the Snort 2.x releases.

The list of inspectors available in Snort 3 includes the following:

- **ARP Spoof inspector**: This inspector module analyzes the **Address Resolution Protocol** (**ARP**) packets and detects various attacks including but not limited to ARP spoofing.

- **CIP inspector**: The **Common Industrial Protocol** (**CIP**) is a protocol used for industrial automation applications and is often used in SCADA networks. The CIP inspector analyzes the CIP messages so as to enable the detection of any attacks.

- **DCE SMB inspector: Distributed Computing Environment/Remote Procedure Calls (DCE/RPC)** is a remote procedure call system. It can work over various layers including SMB, TCP, UDP, and so on. The DCE SMB inspector analyzes DCE/RPC over SMB.

- **DCE TCP inspector**: The DCE TCP inspector analyzes the DCE/RPC protocol over TCP.

- **DNP3 inspector**: The DNP3 inspector decodes and analyzes the DNP3 protocol, which is one of the SCADA-related protocols used with industrial and power automation systems.

- **FTP client inspector: File Transfer Protocol (FTP)** is a protocol for transferring files from a server to a client. FTP works over TCP and usually operates on TCP ports 20 and 21. The FTP client inspector analyzes the FTP client commands and works in conjunction with the FTP server inspector.

- **FTP server inspector**: The FTP server inspector analyzes the FTP server response traffic and works in conjunction with the FTP client inspector.

- **GTP Inspect inspector**: The GTP inspector analyzes the **GPRS Tunneling Protocol (GTP)** traffic.

- **HTTP Inspect inspector**: This is a revamp of the HTTP Inspect module in Snort 2. The HTTP Inspect inspector analyzes HTTP and enables the detection of attacks over web traffic.

- **IEC104 inspector**: This module analyzes IEC104, which is yet another of the SCADA protocols that is used with power systems.

- **IMAP inspector**: The IMAP inspector decodes and analyzes the **Internet Message Application Protocol (IMAP)**, which is used by email clients to retrieve email messages from a remote IMAP server. IMAP uses TCP port 143 and TCP port 993 (encrypted).

- **Modbus inspector**: Modbus is a data communication protocol originally used with **Programmable Logic Controllers (PLCs)**, and nowadays, is often used for communication with industrial electronics devices. The Modbus inspector decodes and analyzes the Modbus protocol traffic. Modbus also belongs to the SCADA suite of protocols.

- **Normalizer inspector**: The normalizer inspector is relevant in an inline mode of operation. This module normalizes the network traffic in order to remove any ambiguity and thereby mitigate evasion attacks. A typical example would be a TCP segment overlap scenario; in this case, the normalizer would edit such that there is no overlapping TCP data.

- **POP inspector**: The POP inspector decodes and analyzes the **Post Office Protocol version 3 (POP3)**, which is used by email clients to retrieve email messages from a remote server.

- **Port Scan inspector**: The Port Scan inspector detects the various scanning attempts over various protocols including TCP, UDP, and ICMP.

- **S7CommPlus inspector**: S7Comm Plus is a proprietary protocol developed by Siemens for use with PLCs. S7Comm Plus is also a SCADA protocol. This inspector inspects and decodes this custom protocol.

- **SIP inspector**: The **Session Initiation Protocol** (**SIP**) is a signaling protocol that enables **Voice over Internet Protocol** (**VoIP**). The SIP inspector decodes and analyzes SIP.

- **SMTP inspector: Simple Mail Transfer Protocol** (**SMTP**) is an internet standard communication protocol for electronic mail transmission. This is one of the common protocols associated with email transmission. SMTP usually operates over TCP port 25. The SMTP inspector decodes and analyzes the SMTP protocol.

- **SSH inspector**: The SSH inspector analyzes the **Secure Shell** (**SSH**) protocol.

- **Stream ICMP inspector**: The Stream ICMP inspector analyzes ICMP flows. This enables the analysis and detection engine to connect and correlate various ICMP packets.

- **Stream IP inspector**: **Internet Protocol** (**IP**) is a connectionless protocol. This inspector tracks IP (network) flows and enables session and flow tracking.

- **Stream TCP inspector**: TCP is a connection-oriented protocol. The Stream TCP inspector is the successor of the various TCP state tracking and reassembly modules implemented in Snort. This module performs state tracking of TCP connections, which is the basis of the analysis of higher-level protocols.

- **Stream UDP inspector**: UDP is a connectionless layer 4 protocol. The Stream UDP inspector defines a UDP flow and enables flow analysis of various UDP pseudo connections.

- **Telnet inspector**: The Telnet inspector decodes and analyzes the Telnet protocol (usually over TCP port 23).

These inspectors perform some of the vital network traffic analysis so that the detection engines can accurately match against Snort signatures. Next, we will look at Snort's detection engine module.

Detection or rule engine

The Detection layer (Rules engine) implements the various detection capabilities, which are pluggable modules in Snort 3. As in Snort 2, the Detect module implements a **multi-pattern search engine** (**MPSE**), a feature essentially made for performance improvement.

One of the classes of pluggable modules is called `IpsOptions`. The `IpsOptions` class implements all of the detection options that are not dependent on inspectors:

Content	Byte_jump	Byte_math
Byte_test	File_data	File_type
Dsize	Pkt_data	Raw_data
Regex	Ttl	tos

Table 3.1 – A few of the rule options that are implemented as the IpsOptions class

Configuration module

Several parameters and settings control the IDS/IPS's functionality and operation. These are usually specified in configuration files and command-line parameters. The configuration module reads the configuration file and command-line parameters and configures the IDS/IPS for correct functioning.

Alerting and logging module

The Snort rule can have several actions: `alert`, `log`, `pass`, and `drop`. When the `alert` and `drop` rules are triggered, it results in an alert. The format and specifics of the generated alert depend on the Snort configuration pertaining to alerting. This module was called the output plugin in the Snort 2.x world.

Snort 3 system architecture

In this section, we will examine how the various key components link to each other and interact to form the Snort 3 system. We will also examine the typical flow of packet processing and the multithreaded approach that is new in Snort 3.

Multithreading

This is one of the critical changes in Snort 3. Based on the number of CPU cores available, the Snort threads are created. The supporting DAQ modules split the network traffic and provide the packets to each Snort thread as required. (The DAQ module may employ techniques such as flow pinning and six-tuple hashing for this purpose.)

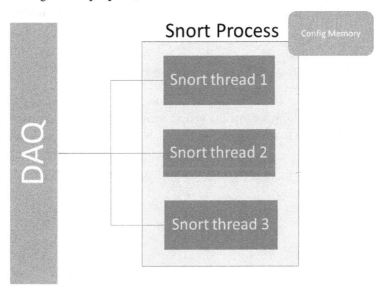

Figure 3.2 – A system with a single Snort process and three processing threads

Figure 3.2 shows a system with a single Snort process and three processing threads. All the threads share the config memory; there is no memory duplication.

In the next section, we will look at the packet flow within each Snort thread.

Packet analysis flow within each Snort thread

At a high level, the packet analysis flow within each Snort 3 thread is like that in Snort 2, namely, stateless decoding, followed by analysis by a set of preprocessors, followed by various detection plugins, and finally, the log/alerting module makes an alert/log as appropriate.

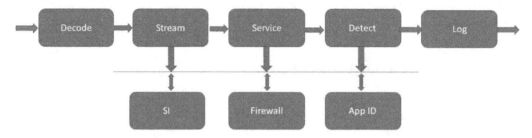

Figure 3.3 – Packet flow through Snort 3

In the previous section, we discussed the various vital components. Now, we will discuss these components from a system architecture perspective. *Figure 3.3* shows the processing flow within each Snort thread. The various modules (such as inspectors) use a just-in-time approach, which means that the relevant processing is not done unless there is a need. For example, HTTP request normalization will not be done unless it is needed for any module. This is done using the **publisher-subscriber (pub-sub)** model.

Summary

In this chapter, we learned about the design goals for Snort 3 and briefly discussed the important components of the system. Finally, we saw the system architecture used by Snort 3 tying all the components together. In the next chapter, we will learn about the basics of installing the Snort 3 system.

4
Installing Snort 3

This chapter will discuss how to install the Snort 3 **intrusion prevention system** (**IPS**). The installation process includes installing all the dependencies (other programs and libraries that Snort will need at runtime) and downloading, compiling, and installing the Snort 3 program. After installation, a configuration step enables Snort to work in a particular environment, load the relevant (specified) modules and rules, and do its job. We will look at configuration in the next chapter.

The Snort IPS can be built for and run on multiple operating systems. In this chapter, we will focus on the installation of Snort 3 on two Linux-based systems, specifically, CentOS and Kali. However, the principles of the Snort 3 installation discussed here are applicable to most operating systems. Therefore, you will be able to follow and apply the same principles to install Snort 3 on most operating systems of your choice.

In this chapter, we will discuss the following topics:

- Choosing an OS for installing Snort 3
- Snort 3 installation process
- Installing Snort 3 on CentOS
- Installing Snort 3 on Kali

Choosing an OS for installing Snort 3

The Snort website lists a set of supported operating systems at the following link: `https://www.snort.org/documents/snort-supported-oss`. The list includes Alpine, CentOS, Debian, Fedora Core, FreeBSD, OpenSUSE, RHEL, Slackware, and Ubuntu.

There are over 600 different Linux distros. Among these, Debian and CentOS are the most popular. For this chapter, we chose Kali, a Debian-based distro, and CentOS.

However, the approach we follow for Snort installation on these two systems generally applies to any system. Therefore, you may follow the process in general and use it on a different Linux distro (for example, Ubuntu) and expect a successful result.

Snort 3 installation process

In this section, we will describe the installation process. Since the process is generic, it may be applied to a different OS distribution. Here is a broad-stroke outline of the installation process. Firstly, the system has to be prepared for the installation, followed by the installation of the required packages and other dependencies, followed by the Snort installation step.

Preparing the system

For each of the cases (CentOS and Kali), we could start with a fresh installation of the operating system or use an already installed system. It is recommended that the latest or a recent version of the distro be used.

Once the operating system is ready, we will update it using the respective package management software. In CentOS, we will use the `dnf` program, and in Kali, we will use `apt`.

Installing dependencies

A set of required packages and libraries must be installed before we can build and install Snort 3. Both CentOS and Kali (Debian) distributions come with their own package management programs, making this step relatively easy and seamless.

In addition, a set of optional packages or libraries may be installed.

Firstly, let us look at the set of libraries and packages listed in the following table:

Package or Library Name	Description
PCRE library	Snort uses this library for regular expression-based pattern matching.
OpenSSL library	Snort has an SSL module. It uses the OpenSSL library for that module. In addition, it uses the library in conjunction with a rule option (protected content) and also for MD5 and SHA hash calculation.
zlib library	This library is used for compression and decompression.
PCAP library	This library is used by the DAQ functionality to capture network traffic for analysis.
LuaJIT library	This library is used for Snort configuration.

Table 4.1 – The list of required packages for Snort 3

All the dependencies (packages) can be installed using the package manager software that is available on the selected OS – for example, `apt` on Debian (Kali) and `dnf` on CentOS.

Installing Snort 3

Once the OS distro is prepared and the dependencies are installed, the next step is to install Snort. We will be building Snort from the source. To do that, we have the following options to get the source code:

1. **Download Snort from the Snort website**: The link for downloading Snort is `https://www.snort.org/downloads`.

    We will download the LibDAQ package and the Snort package. Please note that the versions listed are the latest. You may be downloading a different version based on the status at that time:

    - `snort3-3.1.75.0.tar.gz`

    - `libdaq-3.0.13.tar.gz`

    The files use the `.tar.gz` extensions, meaning they have been archived using the `tar` utility and been compressed using `gzip`. The command-line tool, `tar`, is available natively on all the Linux platforms and may be used to extract the packages, as shown next. Note that we change the directory to a special directory (`~/Snort3`) to extract all the source code. For the `tar` command, you will need to specify the path to the `.tar` file based on where you have downloaded it, for example, `~/Downloads/`:

    ```
    cd ~/Snort3
    tar -zxf ~/Downloads/snort3-3.1.75.0.tar.gz
    ```

    Similarly, extract the other packages as well:

    ```
    cd ~/Snort3
    tar -zxf ~/Downloads/libdaq-3.0.13.tar.gz
    ```

    At this point, we have all the relevant code downloaded, extracted, and ready to install. In the next step, we will see how to get the source code from Git for installation.

2. **Download Snort using Git**: Git is a version control program often used to contribute, manage, and download the managed software. The advantage of using Git is that we can get the latest version of the code at any time.

    From an installation standpoint, the steps are identical after the source is downloaded.

In the next section, we will go through the installation of Snort 3 on CentOS.

Installing Snort 3 on CentOS

This section will describe the steps involved in following the installation process we described in the earlier section on the CentOS system. The default package manager on CentOS is yum, which is being replaced by its successor, dnf. dnf is backward compatible with yum. For our purposes, we used dnf.

Preparing the system

We installed CentOS 7 using the ISO package from the CentOS public repository: https://buildlogs.centos.org/centos/7/isos/x86_64/.

After installation, we installed epel-release on the CentOS system. epel-release stands for **Extra Packages for Enterprise Linux** (**EPEL**) release. EPEL is a software repository that has a wide range of applications that CentOS can use. This will help us prepare the system such that all the required packages and other dependencies for Snort can be installed.

The commands to install the epel-release package and upgrade the system to the latest version is as follows:

```
sudo dnf install -y epel-release
sudo dnf upgrade -y
sudo reboot now
```

The commands need admin privileges and, therefore, the sudo command was used.

Installing build tools

Now, we will install the required build tools so that we can compile and build Snort and DAQ from the source code. The list of packages we need includes gcc, gcc-c++, make, cmake, automake, autoconf, libtool, unzip, flex, bison, curl, and pkgconf:

```
sudo dnf install -y flex bison gcc gcc-c++ make cmake automake
autoconf libtool curl pkgconf
```

Snort 3 code is written in C/C++ and the preceding programs and tools are necessary to compile and build the Snort 3 binary.

Installing dependencies

In this section, we will go through the installation of all the required and optional dependencies. These are listed in *Table 4.1*:

```
sudo dnf install -y libpcap-devel
```

The output of this command is as follows:

```
[osboxes@osboxes ~]$ sudo dnf install -y libpcap-devel
Last metadata expiration check: 0:02:50 ago on Tue 28 Nov 2023 09:29:47 PM EST.
Dependencies resolved.
================================================================================
 Package            Architecture      Version             Repository      Size
================================================================================
Installing:
 libpcap-devel      x86_64            14:1.10.0-4.el9      appstream       160 k

Transaction Summary
================================================================================
Install  1 Package

Total download size: 160 k
Installed size: 241 k
Downloading Packages:
libpcap-devel-1.10.0-4.el9.x86_64.rpm            625 kB/s | 160 kB     00:00
--------------------------------------------------------------------------------
Total                                            296 kB/s | 160 kB     00:00
Running transaction check
Transaction check succeeded.
Running transaction test
Transaction test succeeded.
Running transaction
  Preparing        :                                                      1/1
  Installing       : libpcap-devel-14:1.10.0-4.el9.x86_64                 1/1
  Running scriptlet: libpcap-devel-14:1.10.0-4.el9.x86_64                 1/1
  Verifying        : libpcap-devel-14:1.10.0-4.el9.x86_64                 1/1

Installed:
  libpcap-devel-14:1.10.0-4.el9.x86_64

Complete!
```

Figure 4.1 – Installation of the libpcap-devel package using dnf

Similarly, we install the other packages using the following command:

```
sudo dnf install -y pcre-devel libdnet-devel hwloc-devel openssl-devel
zlib-devel luajit-devel libmnl-devel
```

We can also install selectively from the optional set of packages, as follows:

```
sudo dnf install -y libmnl-devel unwind-devel xz-devel libuuid-devel
hyperscan hyperscan-devel gperftools-devel
```

We will install LibDAQ, a required dependency. The source code for that will be downloaded and installed in the next step (described in the next section). LibDAQ has many modules, including **Netfilter Queue** (**NFQ**). In order to support NFQ, we will need to install the *Netfilter Netlink* and *Netfilter Queue* packages:

```
sudo dnf install -y libnfnetlink-devel libnetfilter_queue-devel
```

Most installed packages have a -devel suffix in their names. For example, in the packages repository, we will have libpcap and libpcap-devel. The devel package contains the files that are required for compiling some code (in our case, Snort) against this library. Otherwise, we only need the regular package (without the -devel suffix), which contains the .so and/or .a library files.

Installing Snort 3

For CentOS Snort installation, we will follow the first approach for getting the source, which is downloading it from www.snort.org/downloads.

LibDAQ

Once the OS distro is prepared and the dependencies are installed, the next step is to install Snort. We will be building Snort from the source. We discussed the two options to get the source for Snort, LibDAQ, and Snort extra packages.

The commands to extract the LibDAQ source package is shown as follows:

```
cd ~/Snort3
tar -zxf ~/sources/libdaq-3.0.13.tar.gz
```

We created a specific directory to extract all the Snort 3-related source code. We named the directory Snort3. The preceding commands initially change the directory to the Snort3 directory and extract the source code for LibDAQ. Let us list the files and directories:

```
cd libdaq-3.0.13
ls
```

There is a README file that explains the steps on how to install the LibDAQ module. We will follow the recommended steps listed, as follows:

```
./bootstrap
./configure
```

Running the bootstrap command creates the configure file, which we run as the next step. This, in turn, creates the Makefile that we need to run the make command:

```
make
sudo make install
```

The make command creates all the binaries and the libraries, and the make install command places them in the relevant locations in the system. This completes the installation for LibDAQ.

Snort 3

Next, we proceed to install the Snort program:

```
cd ~/Snort3
tar -zxf ~/sources/snort3-3.1.75.0.tar.gz
```

The preceding command extracts all the files and directories in a directory named snort3-3.1.75.0. We change the directory into that path. We find a README file here as well that has instructions on how to install the program. We follow suit, as follows:

```
./configure_cmake.sh
```

This step is similar to the configure step and creates a build directory. We have to run make install from the build directory:

```
cd build/
sudo make install
```

This step takes quite a bit of time to complete, and this completes the Snort3 installation. Running the snort -V command confirms the same:

```
[osboxes@osboxes build]$ snort -V

        ,,_      -*> Snort++ <*-
       o"  )~    Version 3.1.75.0
        ''''     By Martin Roesch & The Snort Team
                 http://snort.org/contact#team
                 Copyright (C) 2014-2023 Cisco and/or its affiliates. All rights reserved.
                 Copyright (C) 1998-2013 Sourcefire, Inc., et al.
                 Using DAQ version 3.0.13
                 Using LuaJIT version 2.1.0-beta3
                 Using OpenSSL 3.0.7 1 Nov 2022
                 Using libpcap version 1.10.0 (with TPACKET_V3)
                 Using PCRE version 8.44 2020-02-12
                 Using ZLIB version 1.2.11
                 Using Hyperscan version 5.4.1 2023-04-14
                 Using LZMA version 5.2.5
```

Figure 4.2 – Executing the snort -V command to check installation

That concludes the Snort 3 installation on the CentOS system. We can see that the snort -V command executes and prints out the version of Snort.

Installing Snort 3 on Kali (Debian)

This section will describe the steps involved in preparing the system and setting up Snort on the Kali (Debian-based) system. The default package manager on Kali (Debian) is apt. **Advanced Package Tool (APT)** is the primary package manager for Debian and all related distros, including Kali. We will be using apt to prepare the system and install all dependencies.

Preparing the system

After the Kali system is installed, we will use `apt` to get the system up to date:

```
sudo apt update
sudo apt upgrade
```

After the update and upgrade of the Kali system, the system version is as follows:

```
uname -r
6.5.0-kali3-amd64
```

The `uname` command is used to print out system information.

This is followed by the installation of the required packages and dependencies, which is described in the next section.

Installing dependencies

All the required packages are installed using the `apt` command, as follows:

```
sudo apt install build-essential libpcap-dev libpcre3-dev libnet1-dev
zlib1g-dev luajit hwloc libdumbnet-dev bison flex liblzma-dev openssl
libssl-dev pkg-config libhwloc-dev cmake cpputest libsqlite3-dev uuid-
dev libcmocka-dev libnetfilter-queue-dev libmnl-dev autotools-dev
libluajit-5.1-dev libunwind-dev
```

After these installations are complete, the system is rebooted. There is an additional step that's required so that Snort will find the required libraries at runtime. Two library paths, `/usr/local/lib` and `/usr/local/lib64`, have to be included in `/etc/ld.so.conf.d/local.conf`, and the `ldconfig` command has to be run so that Snort will find the installed LibDAQ library at runtime:

```
#cat /etc/ld.so.conf.d/local.conf

/usr/local/lib
/usr/local/lib64
```

In the next section, we will look at the installation of LibDAQ, followed by the installation of Snort 3. We will be pulling the source code for both using Git.

Installing Snort 3

For Kali (Debian) Snort installation, we will follow the second approach – namely, to get the source using Git. Firstly, a directory was created for downloading the source – namely, `snort-source-files`:

```
mkdir snort-source-files
cd snort-source-files
```

We created a specific directory to extract all the Snort 3-related source code and named the directory `snort-source-files`. We now change the directory into the newly created directory, and the LibDAQ code is downloaded using `git` and installed as follows:

```
git clone https://github.com/snort3/libdaq.git
cd libdaq
./bootstrap
./configure
make
sudo make install
```

Following that, the source for Snort was downloaded using `git`:

```
git clone https://github.com/snort3/snort3.git
```

Here's the output:

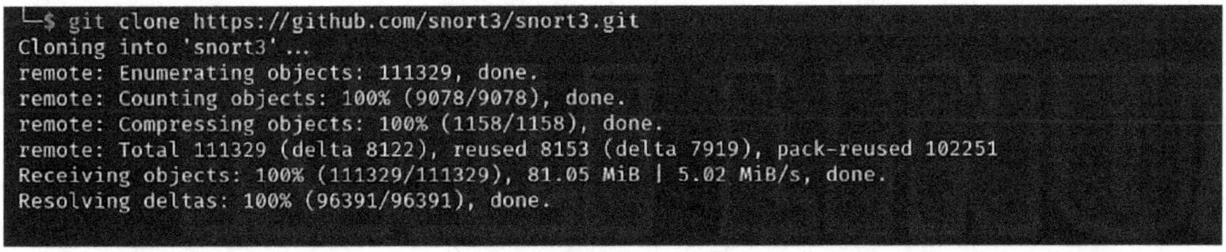

```
└─$ git clone https://github.com/snort3/snort3.git
Cloning into 'snort3' ...
remote: Enumerating objects: 111329, done.
remote: Counting objects: 100% (9078/9078), done.
remote: Compressing objects: 100% (1158/1158), done.
remote: Total 111329 (delta 8122), reused 8153 (delta 7919), pack-reused 102251
Receiving objects: 100% (111329/111329), 81.05 MiB | 5.02 MiB/s, done.
Resolving deltas: 100% (96391/96391), done.
```

Figure 4.3 – Using Git to pull the Snort 3 source code

After the source was downloaded, the installation steps were followed as per the README included in the package:

```
cd snort3
 ./configure_cmake.sh
cd build
make
sudo make install
```

We verify that the installation is complete by running `snort  -V`:

```
┌──(kali⊛kali)-[~/snort-source-files/snort3/build]
└─$ /usr/local/snort/bin/snort -V

   ,,_        -*> Snort++ <*-
  o"  )~      Version 3.1.75.0
   ''''       By Martin Roesch & The Snort Team
              http://snort.org/contact#team
              Copyright (C) 2014-2023 Cisco and/or its affiliates. All rights reserved.
              Copyright (C) 1998-2013 Sourcefire, Inc., et al.
              Using DAQ version 3.0.13
              Using LuaJIT version 2.1.0-beta3
              Using OpenSSL 3.0.11 19 Sep 2023
              Using libpcap version 1.10.4 (with TPACKET_V3)
              Using PCRE version 8.39 2016-06-14
              Using ZLIB version 1.2.13
              Using LZMA version 5.4.5


┌──(kali⊛kali)-[~/snort-source-files/snort3/build]
└─$
```

Figure 4.4 – Executing the snort -V command to check installation

That completes the Snort installation on the Kali (Debian) system as well.

Summary

In this chapter, we walked through the Snort installation process on two different and widely popular operating systems – CentOS and Kali. We looked at the preparation of the system, the installation of dependencies, and finally, the installation of Snort. In the next chapter, we will learn how to configure Snort to set up the IPS.

5
Configuring Snort 3

It's all in the configuration! Imagine a complex machine that can do several complicated tasks, and it has several knobs and switches that control its working. Then, the configuration of this machine is the combination of settings for the knobs and switches. Similarly, Snort 3 is a complex software that has several settings and parameters that determine its working, and much of this working is controlled by its configuration. The term configuration means the combination of values for these settings and parameters. The configuration will determine whether it will perform the analysis of a particular protocol; it will also determine which rules are applied to detect malicious attacks. In short, doing the configuration correctly is critical to getting the best out of Snort 3.

In this chapter, we will discuss the following topics:

- Configuring Snort 3 – how?
- Configuring Snort 3 – what?
- Configuring your environment
- Optimal configuration and tuning
- Managing multiple policies and configurations

Configuring Snort 3 – how?

To get the most out of Snort 3, one must configure it correctly. The right configuration will provide the best performance and best detection rates. In this section, we will look at how Snort configuration is done.

Snort 3 configuration is done mainly via the following:

- Command-line arguments.
- Configuration files.

Command-line arguments

Snort is a system that is written using the C/C++ language. It is possible to pass values to certain Snort variables from the command line. These are called command-line arguments or parameters.

Let us look at one example of how Snort is executed on the command line:

```
snort -V
```

Here, Snort is run with a single command-line parameter, namely -V, which prints out the Snort version. Let us look at another example:

```
snort -c /usr/local/etc/snort/snort.lua -R /usr/local/etc/rules/local.
rules -i ens3 -A alert_fast -s 65535 -k none
```

Here, we see the Snort program being executed with several command-line arguments.

We saw that some command-line parameters, such as -V, are command-line arguments without an associated value. In the preceding example, we see other command-line arguments that go with a value, for example, -c /usr/local/etc/snort/snort.lua. The -c part specifies the configuration file and is followed by the string specifying the configuration file location.

Let us take a quick look at the various command-line options as supported by Snort 3. We will start by looking at the various command-line options related to the help functionality, that is, options that provide information.

Command-Line Argument	Description of the Command-Line Argument
-? <option prefix>	Command-line option for quick help (same as --help-options) (optional). Without the optional argument, that is, -?, it will print out all the various command-line arguments and their meaning.
-h	-h shows help overview (same as --help).
--help-commands [<module prefix>]	This is the command-line option to list all the matching commands. Without the optional prefix, the command-line option will list all the commands. If the prefix is provided, only the commands with the specified prefix will be listed.
--help-config [<module prefix>]	This is the command-line option to list all the matching config options. Without the optional prefix, the command-line option will list all the config options. If the prefix is provided, only the config options with the specified prefix will be listed.

Command-Line Argument	Description of the Command-Line Argument
`--help-counts [<module prefix>]`	This is the command-line option to list all the matching pegs. Without the optional prefix, the command-line option will list all the pegs. If the prefix is provided, only the pegs with the specified prefix will be listed.
`--help-limits`	Prints the upper bounds denoted by `max*`.
`--help-module <module>`	This command-line option prints out the description of the specified module.
`--help-modules`	Lists all available modules with very minimal description.
`--help-modules-json`	Dumps descriptions of all available modules in JSON format.
`--help-options [<option prefix>]`	This is the command-line option to list all the matching command-line options. Without the optional prefix, the command-line option will list all the available options. If the prefix is provided, only the command-line options with the specified prefix will be listed.
`--help-plugins`	Lists all available plugins with brief help.

Table 5.1 – Description of different command-line options

Several of the command-line arguments that Snort 2 supported are still supported by Snort 3 and function in the same way, however, the meanings have changed or become obsolete for others. Let us take a look at a few of the command-line arguments that are carried over from Snort 2:

Argument	Description of the Command-Line Argument
`-c`	Used to specify the configuration file to be used.
`-i`	This option is used to specify the relevant interface for packet acquisition purposes. For example, `-i eth0` (if the traffic on the network interface named `eth0` is to be monitored).
`-R`	Used to specify a rules file to be included in the default policy. The usual way to include the rules file is via config files. The `-R` option is used as a quick way to include an additional rules file without having to edit the configuration. **Note**: In Snort 2, `-R` is used to specify an integer (ID) to be used in the naming of the PID file in the format `snort_intf<ID>.pid`.

Argument	Description of the Command-Line Argument
-k	By default, Snort verifies the header checksum for IP, TCP, UDP, and ICMP. This option is used to make any changes to the default behavior. The following are examples: -k none: Specifies not to validate any header checksum -k notcp: Specifies to validate all but TCP header checksums By not verifying checksums, the system performance may improve; however, the system will be vulnerable to various evasive attacks. In addition, -k none (disabling of checksum validation) is useful in a testing environment when you may be editing the PCAP data but not correcting the packet header checksum.
-T	This option is used to verify or validate the Snort configuration. This command-line parameter does not go along with a value. When this option is specified, Snort will validate the specified command-line parameters and the specified configuration, including the rules, by completing the startup routine and then exiting. If there are any errors during the startup process, this would identify them.
-r	Used to specify a packet capture file for Snort analysis.
-Q	This option denotes that Snort is running in inline mode.

Table 5.2 – Command-line options carried over to Snort 3 from Snort 2

Let's also briefly look at the command-line arguments that are related to **Data Acquisition (DAQ)**:

Argument	Description of the Command-Line Argument
--daq <type>	Option to select packet acquisition module (default is pcap). For example, –daq nfq.
--daq-batch-size	This option sets the size of the batch received for processing.
--daq-dir <dir>	This option is used to specify the location of DAQ library files.
--daq-list	This option will list out the available DAQ choices.
--daq-mode <mode>	Use this option to select the DAQ module operating mode, for example, passive, inline, or read-file. This setting overrides the automatic selection.
--daq-var <name=value>	This option is used to set DAQ-related variables in a name=value format.

Table 5.3 – Descriptions of different command-line options related to DAQ

Since PCAP is a common DAQ, there are a bunch of command-line parameters related to PCAP:

Argument	Description of the Command-Line Argument
`--pcap-file <file>`	Specifies the file that lists out a number of PCAP files to be read (analyzed) by Snort.
`--pcap-list <list>`	Specifies a list of PCAP filenames – separated by a space – to be read (analyzed) by Snort.
`--pcap-dir <dir>`	Specifies the directory for PCAP files.
`--pcap-filter <filter>`	Used to specify a BPF filter (commonly known as a tcpdump filter) that is used to filter out packets.
`--pcap-loop <count>`	Can be used to repeatedly read and process the specified PCAP file in a loop. If the specified count is 0, then loop infinitely. This is an option that may be useful in testing.
`--pcap-no-filter`	Resets to use no filter when reading packets from PCAP files.
`--pcap-show`	Prints a line saying what PCAP file is currently being read.

Table 5.4 – Descriptions of different command-line options related to PCAP

Now, let us look at the command-line options that are related to Snort alerts and logs. Snort supports generating alerts in various formats (for example, `text`, `csv`, and so on) and there needs to be a specified location where these alert files are to be created:

Argument	Description of the Command-Line Argument
`-A <mode>`	This option specifies the mode of alerting. The currently supported modes are `csv`, `fast`, `full`, and `none`. In the `fast` mode, the alert will contain only a subset of the fields or information. The `full` mode will log all the relevant fields to the alert. If the `csv` mode is specified, an alert will be logged in CSV format. Finally, the `none` mode disables the alerting feature. Note: The mode of alert can also be configured by the configuration file.
`-B <mask>`	In certain scenarios, there may be restrictions to logging the source and destination IP addresses (for privacy or other reasons), and in such cases, this option will be handy. The option takes a `mask` input, which will be used to obfuscate the logged IP addresses in alerts and packet dumps.

Argument	Description of the Command-Line Argument
-G	This option is used to specify an ID to be used as a log identifier to uniquely identify events for multiple Snort instances. This option is the same as --log-id.
-v	This option specifies Snort to be more verbose in the logging function.
-U	This option specifies to use UTC for all timestamps
-M	This specifies to log messages (log, warning, and error messages) to syslog.
-f	This disables calling fflush() after logging packets
-L	-L <mode>. The -L option is used to specify the logging mode (none, dump, pcap, or log_*).
-l	This option specifies the directory to be used for logging.
-q	Snort will be on silent mode (no output to stdout).
-O	This is specified in order to obfuscate IP addresses in alerts and logs.

Table 5.5 – Descriptions of different command-line options

Some of the command-line arguments are vestigial ones from Snort's initial days, and they seem more appropriate for a tool such as tcpdump. Let us take a quick look at these:

Argument	Description of the Command-Line Argument
-d	This option will cause Snort to dump the application later.
-e	The -e option tells Snort to display the layer 2 header information. (This is an option present in the tcpdump tool.)
-C	This option will cause Snort to print out payloads with character data only.
-n <count>	The -n option is used to inform Snort to stop after receiving count number of packets. (This is a behavior present in the tcpdump tool.)
-s <snaplen>	This option is used to specify the length of the packet to be retrieved for analysis. The --snaplen option behaves exactly like the -s option.
-X	This option specifies the program to dump the raw (hex) packet data at the link layer.

Table 5.6 – Descriptions of different command-line options

There is a set of command-line options that deals with dumping various kinds of information. Here are those options:

Argument	Description of the Command-Line Argument
`--dump-builtin-options`	Additional options to include with `--dump-builtin-rules` stubs
`--dump-builtin-rules [<module prefix>]`	Output stub rules for selected modules (optional)
`--dump-config`	Dump config in JSON format (all | top). If *top* is specified only, the main policy config is listed. If *all* is specified, all the policy configs will be listed.
`--dump-config-text`	Dump config in text format
`--dump-dynamic-rules`	Output stub rules for all loaded rules libraries
`--dump-defaults [<module prefix>]`	Output module defaults in Lua format (optional)
`--dump-rule-databases`	Dump rule databases to the given directory.
`--dump-rule-deps`	Dump rule dependencies in JSON format for use by other tools
`--dump-rule-meta`	Dump configured rule info in JSON format for use by other tools
`--dump-rule-state`	Dump configured rule state in JSON format for use by other tools
`--dump-version`	Output the version, the whole version, and only the version

Table 5.7 – Descriptions of different command-line options

In this section, we discussed Snort's various command-line parameters and saw how they control its runtime behavior. We discussed the command-line parameters that Snort 3 inherited from Snort 2 as well as the new ones that were introduced in Snort 3. In the next section, let us look at the Snort configuration file and its importance.

Configuration files

A configuration file contains most, if not all, of Snort's operational parameters and settings. In Snort 3, the configuration file is not a simple text file but is specified using Lua, a scripting language. An important use of Lua is as a configuration language, and it lends itself well to being embedded into other applications.

There is no standard name for the Snort configuration file, although snort.lua is a very commonly used name. Irrespective of the name, the configuration file that is provided using the -c command-line argument is the configuration file:

```
snort -c snort.lua --daq pcap -r ~/Downloads/PCAPS/http.cap
```

In the preceding command, Snort is started with snort.lua as the configuration file, and PCAP as the DAQ method. The command also specifies (using the -r option) a PCAP file called http.pcap which is the input packet capture file to be read and analyzed.

If you download the Snort source package, you will find a set of predefined configuration files that may be used as a starting point:

```
ls snort3-3.1.47.0/lua/
balanced.lua   connectivity.lua   inline.lua   security.lua      snort_
defaults.lua   talos.lua    max_detect.lua    snort.lua
```

Please note that all these config files are not the same. This will be explained in the next section.

Configuration defaults and tweaks

If we look in snort.lua, we will see that it includes a default configuration file called snort_defaults.lua, followed by specific configuration for various modules.

For Snort's internal variables, there are default values. It can be internal defaults, which means the value is set to a particular value within the Snort binary, or it can be an external default, which means that the variable is set to a particular value in a default configuration file. snort_defaults.lua is a configuration file that contains the external default values. We would include this defaults file in the configuration file, and then change any value that we would like to modify from the default setting.

As we discussed how snort.lua included snort_defaults.lua, it is common for a configuration file to include other files that contain configuration information.

Now, Snort supports another command-line parameter, called –tweaks, which may be used to layer the configuration. In the lua directory listing given previously, we can see the following Lua files – balanced.lua, security.lua, connectivity.lua, and max_detect.lua. Each of these Lua files has some modifications to the snort.lua settings for a reason. We make these tweaks effectively by using the –tweaks command-line parameter as follows:

```
snort -c snort.lua --tweaks security --daq pcap -r ~/Downloads/PCAPS/
http.cap
```

When we have this setting, we notice the following in the Snort startup logs, which explains what is going on:

```
--------------------------------------------------
o")~    Snort++ 3.1.47.0
--------------------------------------------------
Loading snort.lua:
Loading snort_defaults.lua:
Finished snort_defaults.lua:
Loading security.lua:
Finished security.lua:
```

The logs file shows how the configuration files are being processed – the configuration loading starts with snort.lua, then proceeds to snort_defaults.lua since that is being included by snort.lua, and finally, security.lua is processed.

Configuration syntax and parsing

Parsing the configuration file(s) is part of the Snort startup routine. In previous versions (Snort 2.9.x and earlier), the configuration file was specified using custom syntax. This led to a situation where there was no consistency in the specification format, and this even differed between modules. We can see this if we take a quick look at the frag3 preprocessor configuration and compare it to the stream5_global preprocessor configuration in Snort 2.

Firstly, let us look at the frag3 preprocessor configuration. It can be noted that it uses *space* as the delimiter:

```
# Target-based IP defragmentation.  For more information, see README.
frag3
preprocessor frag3_global: max_frags 65536
preprocessor frag3_engine: policy windows detect_anomalies overlap_
limit 10 min_fragment_length 100 timeout 180
```

Now, let us look at the `stream5_global` configuration. We see that it uses a *comma* as the delimiter between parameters:

```
preprocessor stream5_global: track_tcp yes, track_udp yes, track_icmp
no, max_tcp 262144, max_udp 131072, max_active_responses 2, min_
response_seconds 5
```

The syntax and formatting in Snort 2 are not standardized across all modules.

Snort 3 made a significant change regarding this by using Lua to specify the configuration. Lua is essentially a scripting language, and one of its important uses is as a configuration language. The popular NIDS Suricata also uses Lua for its configuration.

Let's look at how we enable two modules in Snort 3 using Lua-based configuration files, and how a parameter in each of these modules is modified:

```
http_inspect = { }
http_inspect.decompress_pdf = true
```

The first line enables the `http_inspect` module, and the second line enables the `decompress_pdf` feature:

```
stream_ip = { }
stream_ip.min_frag_length = 100
```

Here, the first line enables the `stream_ip` module, and the second line sets the `min_frg_length` setting to `100`.

With this new approach (using Lua), there is a standard and consistent way of specifying configuration. As mentioned earlier, the move to use Lua for configuration is significant. However, there is one downside: Snort 3 is not backward compatible with Snort 2 configuration files. To help with this transition, however, Snort 3 comes with a tool called **Snort2Lua**, which can be used to convert Snort 2 configuration and rules to compatible Snort 3 configuration and rules.

Validating a configuration file

Validating a configuration is often a prudent step, especially before running Snort live. Let us discuss a bit about default values. Snort maintains several internal variables, such as timeout values, or a CIDR value that needs to be initialized for its operation. These variables can have a default value that is set within the Snort binary (as hardcoded values in C/C++ code). These are called internal default values. Alternatively, these variables can have external default values, which are set within the configuration file as a default (or predefined) setting. Finally, these values can be set as needed within the configuration file.

Also, Snort variables can be scalar or non-scalar. Scalar variables are those variables that have a single value, for example, variables of the integer, float, or string type. Non-scalar variables are those that can have a set of values, for example, lists, arrays, or dictionaries.

Most scalar variables have internal defaults, whereas non-scalar variables are initialized using external defaults. The external defaults are usually set in a default configuration file, which is included within the main configuration file. For example, the defaults file may be named `snort_defaults.lua` and the main configuration file may be named `snort.lua`. Then, the `snort.lua` file will include the `snort_defaults.lua` file using the include directive.

Running Snort with just the `-c` command-line argument tells Snort to validate the specified configuration and exit:

```
snort -c /usr/local/etc/snort/snort.lua
```

In addition to the configuration, we may also validate a rules file by using the `-R` command-line option:

```
snort -c /usr/local/etc/snort/snort.lua -R /usr/local/etc/rules/local.
rules
```

After the validation is complete, Snort prints out the following to stdout:

```
Snort successfully validated the configuration (with 0 warnings).
```

In this section, we discussed the two main ways of configuring Snort 3, namely via configuration files and command-line arguments. In the next section, we will get into the details of Snort configuration.

Configuring Snort 3 – what?

We saw the ways in which we configure Snort, namely, using command-line arguments and configuration files. In this section, we would like to understand the *what* side of Snort configuration.

Being a modular system was one of the design goals of the Snort 3 system. By this design, Snort is a heavily modularized system. Almost all the modules are plugins as well. The configuration is segregated by modules (or plugins) and thus well organized.

The `snort.lua` configuration file that ships with the Snort package gives a good guideline on how to do the Snort configuration. The steps are as follows:

1. Configure defaults.
2. Configure inspection.
3. Configure bindings.
4. Configure performance.
5. Configure detection.

6. Configure filters.

7. Configure outputs.

Let us walk through each of these sections starting with configuring defaults.

Configuring defaults

`snort.lua` references a file called `snort_defaults.lua`, as shown in the following code snippet. This is a strategy where several Snort configuration parameters are set with default values in the defaults file, which is included in the main configuration file. Thereafter, any parameter that needs to be changed can be specified in the main configuration file:

```
include 'snort_defaults.lua'
```

Now, let us discuss the contents of the default configuration file, namely `snort_defaults.lua`. The Lua files (as opposed to the text config files in the Snort 2 world) are not parsed, but they are scripts that are run.

The default Lua file – `snort_defaults.lua` – creates several default values that are used by the main `snort.lua` configuration file. There are default values for various modules, such as HTTP, FTP, SMTP, and GTP. The script also combines and groups various values to create other non-scalar values to be assigned as defaults. Let us look at one such example.

The defaults file defines three types of variables, namely path, network, and port variables. Subsequently, the script groups these together to create a non-scalar value and assigns it to `default_variables`. Let us look closely at each of these types in the following subsections.

Path variables

These set the location of various files, such as rules files, stub rules, shared object rules, and whitelist or blacklist files. `Path` is a scalar variable; however, it is better to have the defaults for various paths as external defaults than internal defaults:

```
-- Path to your rules files (this can be a relative path)
RULE_PATH = '../rules'
BUILTIN_RULE_PATH = '../builtin_rules'
PLUGIN_RULE_PATH = '../so_rules'

-- If you are using reputation preprocessor set these
WHITE_LIST_PATH = '../lists'
BLACK_LIST_PATH = '../lists'
```

Network variables

These are variables that store individual IP addresses or CIDR values for the relevant servers in the home network. In the default configuration, HOME_NET is defined as any, and therefore, all the servers are also defined as any:

```
-------------------------------------------------------------
-- default networks - used in Talos rules
-------------------------------------------------------------
-- define servers on your network you want to protect
DNS_SERVERS = HOME_NET
FTP_SERVERS = HOME_NET
HTTP_SERVERS = HOME_NET
SIP_SERVERS = HOME_NET
SMTP_SERVERS = HOME_NET
SQL_SERVERS = HOME_NET
SSH_SERVERS = HOME_NET
TELNET_SERVERS = HOME_NET
```

Port variables

These variables correspond to the various protocols that are commonly used in networks and their default values. For example, the HTTP protocol is associated with ports 80, 8080, 3128, and so on. The various port variables and their default values are as follows (taken from snort_defaults.lua):

```
-------------------------------------------------------------
-- default ports
-------------------------------------------------------------
FTP_PORTS = ' 21 2100 3535'
HTTP_PORTS =
[[
    80 81 311 383 591 593 901 1220 1414 1741 1830 2301 2381 2809 3037
3128
    3702 4343 4848 5250 6988 7000 7001 7144 7145 7510 7777 7779 8000
8008
    8014 8028 8080 8085 8088 8090 8118 8123 8180 8181 8243 8280 8300
8800
    8888 8899 9000 9060 9080 9090 9091 9443 9999 11371 34443 34444
41080
    50002 55555
]]
MAIL_PORTS = ' 110 143'
ORACLE_PORTS = ' 1024:'
SIP_PORTS = ' 5060 5061 5600'
```

```
SSH_PORTS = ' 22'
FILE_DATA_PORTS = HTTP_PORTS .. MAIL_PORTS
```

These three sets are then grouped together by `snort_defaults.lua` to create a non-scalar variable called `default_variables` (remember Lua is a scripting language):

```
default_variables =
{
    nets =
    {
        HOME_NET = HOME_NET,
        EXTERNAL_NET = EXTERNAL_NET,
        DNS_SERVERS = DNS_SERVERS,
        FTP_SERVERS = FTP_SERVERS,
        HTTP_SERVERS = HTTP_SERVERS,
        SIP_SERVERS = SIP_SERVERS,
        SMTP_SERVERS = SMTP_SERVERS,
        SQL_SERVERS = SQL_SERVERS,
        SSH_SERVERS = SSH_SERVERS,
        TELNET_SERVERS = TELNET_SERVERS,
    },
    paths =
    {
        RULE_PATH = RULE_PATH,
        BUILTIN_RULE_PATH = BUILTIN_RULE_PATH,
        PLUGIN_RULE_PATH = PLUGIN_RULE_PATH,
        WHITE_LIST_PATH = WHITE_LIST_PATH,
        BLACK_LIST_PATH = BLACK_LIST_PATH,
    },
    ports =
    {
        FTP_PORTS = FTP_PORTS,
        HTTP_PORTS = HTTP_PORTS,
        MAIL_PORTS = MAIL_PORTS,
        ORACLE_PORTS = ORACLE_PORTS,
        SIP_PORTS = SIP_PORTS,
        SSH_PORTS = SSH_PORTS,
        FILE_DATA_PORTS = FILE_DATA_PORTS,
    }
}
```

This is referenced by the `ips` module, as we will see in a later section.

Configuring inspection

This section configures all the modules that are related to inspection. All the inspector modules will be enabled or disabled in this section. Also, the configuration for each of these modules is done here.

The format for enabling any module is as follows, where *mod* is the name of the module being configured:

```
mod = { }
```

This enables the *mod* module and sets its variables to internal defaults.

Alternatively, when a module uses external defaults, the usage will be as follows:

```
mod = default_mod_variables
```

Here, `default_mod_variables` will be set in the default configuration file, for example, `snort_defaults.lua`.

Here is a part of the `Inspection` section in the `snort.lua` file:

```
stream = { }
stream_ip = { }
stream_icmp = { }
stream_udp = { }
stream_user = { }
stream_file = { }
arp_spoof = { }
back_orifice = { }
dns = { }
imap = { }
netflow = {}
normalizer = { }
pop = { }
rpc_decode = { }
sip = { }
ssh = { }
ssl = { }
telnet = { }
..
```

All these modules are enabled with internal defaults. Then, we also have the following:

```
gtp_inspect = default_gtp
port_scan = default_med_port_scan
smtp = default_smtp
ftp_server = default_ftp_server
```

These modules are enabled with external default values (as defined in `snort_defaults.lua`).

Configuring bindings

The configurations pertaining to the binder inspector and the autodetection wizard are done in this section. In short, the binder and wizard modules together try to bind every flow to a certain inspector for inspection. The binder module does this using the help of a set of rules; for example, one rule is as follows:

```
    { when = { proto = 'udp', ports = '53', role='server' },  use = {
type = 'dns' } },
```

The binder module uses this rule and binds a traffic flow to the `dns` inspector when the flow is over the UDP protocol, with the server port equal to 53.

The wizard module analyzes the data over the flow and recognizes keywords; for example, it may recognize the GET method of an HTTP request and assign the flow to the HTTP inspector.

The configuration for the binder module is therefore a set of rules similar to the rule shown previously. These rules will need to be edited as needed.

The configuration for the wizard consists of special keywords and patterns. The default configuration file (`snort_defaults.lua`) defines the common set of methods, requests, and commands of various protocols, such as HTTP, SIP, and Telnet.

The HTTP methods given in `snort_defaults.lua` are as follows:

```
http_methods =
{
    'GET', 'HEAD', 'POST', 'PUT', 'DELETE', 'TRACE', 'CONNECT',
    'VERSION_CONTROL', 'REPORT', 'CHECKOUT', 'CHECKIN', 'UNCHECKOUT',
    'MKWORKSPACE', 'LABEL', 'MERGE', 'BASELINE_CONTROL',
    'MKACTIVITY', 'ORDERPATCH', 'ACL', 'PATCH', 'BIND', 'LINK',
    'MKCALENDAR', 'MKREDIRECTREF', 'REBIND', 'UNBIND', 'UNLINK',
    'UPDATEREDIRECTREF', 'PROPFIND', 'PROPPATCH', 'MKCOL', 'COPY',
    'MOVE', 'LOCK', 'UNLOCK', 'SEARCH', 'BCOPY', 'BDELETE', 'BMOVE',
    'BPROPFIND', 'BPROPPATCH', 'POLL', 'UNSUBSCRIBE', 'X_MS_ENUMATTS',
    'NOTIFY * HTTP/', 'OPTIONS * HTTP/', 'SUBSCRIBE * HTTP/', 'UPDATE
* HTTP/',
    '* * HTTP/'
}
```

Next, we will discuss the performance monitoring and latency enforcing configurations.

Configuring performance

There are use cases for a system where a certain level of performance expectations needs to be met. These expectations are usually high when the system is deployed in an inline fashion and operates as an IPS. When the system's performance degrades, this will result in higher latency for the packets. When packet latency increases, this will lead to reduced network throughput and efficiency.

To ensure a high level of system performance, there are mainly two aspects to consider. Firstly, the system should be able to track a set of performance-related metrics, which can be monitored by admins. Secondly, when certain requirements are not met, the system can be configured to respond in certain ways to mitigate the issue. For example, if the packet latency is greater than a threshold, the packet can be sent by a fast path (skipping the remaining analysis).

There are two relevant Snort modules for performance monitoring and latency in Snort:

- **The perf monitor**: This is an inspector module that keeps a set of metrics that may be used to understand and monitor the system's performance.

- **The latency module**: This is a core module that monitors packet and rule latency and offers certain controls.

There is packet-level granularity as well as rule-level granularity. The `packet.max_time` setting sets a threshold for the maximum processing time for a single packet. If the processing time of any packet (the time taken by the IPS engine to process that packet) exceeds this setting, then the packet is marked as a potential fast-path candidate. This parameter value is specified in microseconds:

```
int latency.packet.max_time = 500
```

Fast path is the term that is used for when the IPS stops analyzing a packet and sends it along – when the processing time exceeds the `packet.max_time` setting. In order for the packet to be sent by fast path, `packet.fastpath` has to be configured as `true`:

```
bool latency.packet.fastpath = false
```

There is also a rule-level granularity for the latency module. `rule.max_time` puts a higher limit on the time spent analyzing any single rule (for any packet). This value is also specified in microseconds:

```
int latency.rule.max_time = 500
```

When the time taken for any rule exceeds `rule.max_time` for a certain number of times as configured by the `rule.suspend_threshold` setting, that rule is marked as expensive and potentially suspended if `rule.suspend` is set as `true`:

```
bool latency.rule.suspend = false
int latency.rule.suspend_threshold = 5
int latency.rule.max_suspend_time = 30000
```

Next, let us discuss the configuration of the detection module.

Configuring detection

This section contains configurations specific to Snort rules, the variables that are needed for the rules, rule actions, rule classifications, and so on. The default values for this section will be present in the default configuration file (for example, `snort_defaults.lua`).

We saw in the *Configuring defaults* section how a non-scalar variable called `default_variables` was created by `snort_defaults.lua`. This is used by the `snort.lua` script as follows:

```
ips =
{
    -- use this to enable decoder and inspector alerts
    enable_builtin_rules = true,

    -- use include for rules files; be sure to set your path
    -- note that rules files can include other rules files
    -- (see also related path vars at the top of snort_defaults.lua)

    variables = default_variables
}
```

In the next section, we will discuss the set of post-detection filters.

Configuring filters

This section of the configuration deals with the set of post-detection filters – suppression filters, rate filters, and event filters. For every alert generated by Snort, there is a corresponding `gid` (generator ID) and `sid` (signature ID). Each Snort module that can generate an alert has a respective `gid` value, and each alert that is generated from every module has a unique `sid` value within that `gid` value. All the post-detection filters use the combination of `gid` and `sid`.

There are three parts to the configuring filters section, namely suppression filters, event filters, and rate filters.

The suppression filters are configured as shown in the following code snippet using the non-scalar variable called `suppress`. The `suppress` variable is a list of records, each denoting a criterion for an event that should be suppressed. Each record can denote a combination of `gid`, `sid`, `track`, and `ip`. The `track` component specifies whether the `ip` that is tracked is the source of the destination IP address:

```
-------------------------------------------------------------
-- 6. configure filters
-------------------------------------------------------------
-- below are examples of filters
-- each table is a list of records
--[[
suppress =
{
    -- don't want to see any of these
    { gid = 1, sid = 1 },

    -- don't want to see anything for a given host
    { track = 'by_dst', ip = '1.2.3.4' }

    -- don't want to see these for a given host
    { gid = 1, sid = 2, track = 'by_dst', ip = '1.2.3.4' },
}
--]]
```

There are situations when the alert volume has to be reduced and not completely suppressed. The event filter section deals with that use case. A non-scalar variable called `event_filter` is used to list all the various records for event filtering. Each of these records specifies a unique case for filtering:

```
--[[
event_filter =
{
    -- reduce the number of events logged for some rules
    { gid = 1, sid = 1, type = 'limit', track = 'by_src', count = 2,
seconds = 10 },
    { gid = 1, sid = 2, type = 'both',  track = 'by_dst', count = 5,
seconds = 60 },
}
--]]
```

There are different types of event filtering, namely limit, threshold, and both. The limit type is used when all the alerts after a certain limit are suppressed for the specified time period. For example, let us discuss the case listed, namely gid = 1, sid = 1, type = 'limit', track = 'by_src', count = 2, seconds = 10. Here, this deals with the alert with gid = 1 and sid = 1. For that particular alert, the filter limits to 2 alerts every 10 seconds from the same source.

The next type is threshold, in which case the alert is suppressed until a certain number of alerts is reached within the specified time period.

The third type is both, where an alert is generated only when the specified count is reached in the specified time period, and also limits the number of alerts generated to the specified count in that time period:

```
--[[
rate_filter =
{
    -- alert on connection attempts from clients in SOME_NET
    { gid = 135, sid = 1, track = 'by_src', count = 5, seconds = 1,
      new_action = 'alert', timeout = 4, apply_to = '[$SOME_NET]' },

    -- alert on connections to servers over threshold
    { gid = 135, sid = 2, track = 'by_dst', count = 29, seconds = 3,
      new_action = 'alert', timeout = 1 },
}
--]]
```

Rate-based filtering is very useful in protecting against DOS-like attacks. In this case, the action taken can be changed when the number of events exceeds a certain threshold as specified by the configuration. This change in action is specified by the new_action component.

In the next section, let us look at how the alert logger module is configured.

Configuring output

When a particular rule is matched, an alert event is triggered. The format for the alert is determined by the configuration. In the configuration file (for example, snort.lua), we specify the alert output format as follows:

```
-- event logging
-- you can enable with defaults from the command line with -A <alert_
type>
-- uncomment below to set non-default configs
--alert_csv = { }
--alert_fast = { }
alert_full = { }
```

```
--alert_sfsocket = { }
--alert_syslog = { }
--unified2 = { }
```

`--` indicates a commented line. In the preceding case, the `alert_full` method for alerting is enabled.

The various formats are as follows:

- `alert_full`: This is the format where all the relevant fields, including the packet's hexdump, are logged as part of the alert. By default, this alert is logged to `stdout`; however, it may be logged to a file based on a configuration parameter (`alert_full.file`). The `limit` parameter can be used to specify a limit to the file size after which the file is rolled over:

  ```
  bool alert_full.file = false
  int alert_full.limit = 0
  ```

- `alert_fast`: This is an option for quick logging rather than detailed logging. Hence, all fields are not logged, and packet data is not logged by default. Optionally, the configuration can be modified so that the module includes packet data as part of the alert. By default, this alert is logged to `stdout`; however, it may be logged to a file (`alert_fast.txt`) based on a configuration parameter (`alert_full.file`). The `limit` parameter can be used to specify a limit to the file size after which the file is rolled over:

  ```
  bool alert_fast.file = false
  bool alert_fast.packet = false
  int alert_fast.limit = 0
  ```

- `alert_csv`: This is an option where the alert details are logged in CSV format. The individual fields that are included in the CSV are based on configuration. By default, this alert is logged to `stdout`; however, it may be logged to a file (`alert_csv.txt`) based on a configuration parameter (`alert_full.file`). In addition, the default field separator value of a comma (`,`) can also be changed with the configuration:

  ```
  bool alert_csv.file = false
  multi alert_csv.fields = 'timestamp pkt_num proto pkt_gen pkt_
  len dir src_ap dst_ap rule action'
  int alert_csv.limit = 0
  string alert_csv.separator = ', '
  ```

The alerting behavior is also affected by the rate and detection filter modules. Since these are closely related to the rule evaluation and rule options, we shall discuss them in *Chapter 13*.

In the next section, let us discuss how anyone can configure the Snort system for their environment.

Configuring your environment

In this section, we will learn how one would configure the Snort system for their own network or environment. As we mentioned earlier, Snort has several configuration parameters, and the effectiveness and efficiency of the system depend on how well the system is configured for the specific environment where it operates.

In this section, we will discuss some of the key configuration parameters that need to be configured for a specific environment. We will start with the section of the configuration file called *Network Variables*. This is the section that defines IP and port variables.

HOME_NET

This is an IP variable that represents a list of IP addresses and subnet values that represent the network that is being protected. The default value for this parameter is any. This must be changed to a list of IP addresses and subnets that most accurately represent your network.

This variable is most often referenced in Snort signatures. When a signature has the destination IP address field set as HOME_NET, it means that the signature should be applied to traffic that is bound to the network that is being protected.

The next variable we look at is called EXTERNAL_NET, which is also an IP list variable.

EXTERNAL_NET

If HOME_NET represents the network that is being guarded or protected, then EXTERNAL_NET represents the outside internet. Hence, it is often set to any. It is also common practice to set this variable as !$HOME_NET, which means all the IP addresses excluding the network that is being protected.

In addition to HOME_NET and EXTERNAL_NET, there are several other IP address variables, such as DNS_SERVERS, FTP_SERVERS, HTTP_SERVERS, and SMTP_SERVERS, which also need to be configured with accurate information as it applies to the environment.

Next, we will look at some of the port variables.

HTTP_PORTS

HTTP_PORTS is a port variable that represents all the TCP ports used for HTTP in the network environment under consideration. By default, this variable is set as follows:

```
HTTP_PORTS =
[[
    80 81 311 383 591 593 901 1220 1414 1741 1830 2301 2381 2809 3037
3128
    3702 4343 4848 5250 6988 7000 7001 7144 7145 7510 7777 7779 8000
```

```
8008
    8014 8028 8080 8085 8088 8090 8118 8123 8180 8181 8243 8280 8300
8800
    8888 8899 9000 9060 9080 9090 9091 9443 9999 11371 34443 34444
41080
    50002 55555
]]
```

As you can see, all the possible TCP ports that are often used for HTTP are listed here. However, this may not be the most accurate setting for your network. So, you need to list all the HTTP ports used in your network.

Similar to HTTP_PORTS, there are other port variables, such as FTP_PORTS, SSH_PORTS, and MAIL_PORTS, that also need to be configured with the most accurate value for your network.

Next, let us look at how one would configure the stream_tcp inspector for their environment.

The stream_tcp inspector

We will discuss the various configuration parameters of the stream_tcp inspector in *Chapter 9*. However, in order to explain the importance of configuring the inspector for a specific environment, we will take the following parameter of the stream_tcp inspector: stream_tcp.policy. The following code snippet shows the listing of the parameter as displayed by the help-module command-line option in Snort (snort --help-module stream_tcp):

```
enum stream_tcp.policy = 'bsd': determines operating system
characteristics like reassembly { 'first' | 'last' | 'linux' | 'old_
linux' | 'bsd' | 'macos' | 'solaris' | 'irix' | 'hpux11' | 'hpux10' |
'windows' | 'win_2003' | 'vista' | 'proxy' | 'asymmetric' }
```

The default setting of policy is bsd, which means that Snort would carry out the TCP stream reassembly operation as done by the BSD operating system. However, if one's network is Windows-based, it would be necessary to set this parameter to windows.

In this section, we discussed the importance of configuring the Snort parameters for the environment where Snort will be run. In the next section, we will discuss the topic of optimal configuration.

Optimal configuration and tuning

In the previous section, we discussed the importance of configuring the Snort system for the network environment where it will be operating. However, the Snort configuration is complex and consists of hundreds of parameters and settings. It is almost impossible to get the optimal configuration setting for any environment on the first attempt. Thus, the process of arriving at the optimal configuration for Snort in any network environment is a process. This process of continually making changes in order to improve the effectiveness and efficiency of the Snort system is called tuning.

In addition, the network itself is not a static object. The network also undergoes changes as time passes. Thus, there is a need to continually tune the Snort configuration to aim for the optimal configuration.

In the next section, we shall discuss the topic of having more than one configuration for Snort.

Managing multiple policies and configurations

We discussed the `stream_tcp.policy` configuration parameter in the *Configuring your environment* section. We saw how the default setting for the parameter was `bsd` and how it should be changed to `windows` when the environment contains Windows machines. However, in reality, we often encounter environments where the network contains a mix of multiple operating systems. In such cases, the `stream_tcp` module must perform TCP stream reassembly operating as Windows does, when the endpoint machine is Windows, and as BSD when the endpoint machine is BSD. This requires Snort to have multiple configurations and policies.

This is made possible using the Binder inspector, which is a special inspector in Snort 3. We will discuss the Binder inspector in *Chapter 8*.

Consider a network where a subset of the network (`192.168.0.0/16`) consists of BSD machines:

```
binder =
{
   { when = { nets = '192.168.0.0/16', proto = 'tcp'},
     use = { file = 'bsd.lua' } },

   -- use the default inspectors:
   -- (similar to a Snort 2 default preprocessor config)
   { when = { proto = 'tcp' }, use = { type = 'stream_tcp' } },
   { when = { service = 'http' }, use = { type = 'http_inspect' } },

   -- figure out which inspector to run automatically:
   { use = { type = 'wizard' } }
}
```

The configuration specifies whether the IP address matches the `192.168.0.0./16` range and the transport protocol is TCP, then uses the `bsd.lua` file for configuration. Within `bsd.lua`, we could set `stream_tcp.policy` to `bsd` and have the default `stream_tcp.policy` as `windows`.

In this section, we discussed how the binder inspector can be used to manage multiple policies and configurations. Next, we will summarize what we discussed and learned in the chapter.

Summary

In this chapter, we discussed and reviewed the Snort 3 configuration aspects. Snort IDS/IPS version 3 is a complex software with several configuration settings and parameters that determine how the network traffic is analyzed to detect malicious attacks, what protocols are analyzed, how alerts are logged, and much more. Understanding and configuring is a critical part of running the Snort IDS/IPS. The effectiveness and performance of the Snort system depend on this aspect.

In the next chapter, we will discuss Snort's DAQ module and the various mechanisms involved in that functionality.

Part 3:
Snort 3 Packet Analysis

The third part of the book discusses packet processing in Snort 3 in detail, including packet acquisition, packet decoding, and the inspector component. This section will provide you with an in-depth understanding of the packet analysis process and how to configure each of the Snort modules.

This part has the following chapters:

- *Chapter 6, Data Acquisition*
- *Chapter 7, Packet Decoding*
- *Chapter 8, Inspectors*
- *Chapter 9, Stream Inspectors*
- *Chapter 10, HTTP Inspector*
- *Chapter 11, DCE/RPC Inspectors*
- *Chapter 12, IP Reputation*

6
Data Acquisition

The **Data Acquisition (DAQ)** module, or layer, deals with packet I/O. Its single purpose is to facilitate the delivery and transmission of network packets to and from Snort. Historically, this functionality was tightly coupled within Snort code, and as Snort grew, there was a need to simplify and abstract it out. The DAQ feature was implemented in the Snort 2.9 release.

In this chapter, we're going to cover the following main topics:

- The functionality of the DAQ layer
- The performance of the DAQ layer
- Packet capture functionality in Snort
- The Snort 3 implementation of the DAQ layer
- Configuring DAQ

The functionality of the DAQ layer

The main functionality of the DAQ layer is to facilitate the delivery of network packets from the network to Snort, and facilitate the transmission of packets back to the network when appropriate. Let's discuss each of these functionalities:

- **Facilitate the delivery of packets from the network to Snort for analysis**:

  This functionality is the most basic feature required for any system that inspects network traffic and performs analysis. For example, it is necessary for programs such as tcpdump and basic intrusion detection systems.

  The DAQ library provides a layer of abstraction for the packet capture-related function calls, which leads to the simplification of code at the Snort level. Snort code does not need to know about the details of any specific packet capture mechanism. The DAQ layer implements a set of necessary functions as an API, which is invoked at the Snort level.

DAQ is implemented as a library component and a module component at the implementation and design levels. Snort code invokes the necessary functions using the API defined in the library. The various DAQ modules implement all the details specific to any packet capture mechanism. This design also supports third parties in developing DAQ modules created for specific needs, such as high-speed capture using a specific network card.

- **Facilitate the transmission of packets to a network:**

  Intrusion detection systems work by monitoring network traffic, and most likely, this monitoring is done passively. However, as IDS evolved, a reactive component was added using a rule option called resp (Snort 1.8 release). This functionality enabled the IDS to respond – for example, by sending TCP Reset (RST) packets to terminate connections. This feature required the ability to transmit packets from the IDS, mainly to thwart a malicious connection. After that, when intrusion prevention systems were created, which operated in an inline fashion, transmission of packets became a necessary option. Let's briefly discuss the two cases, namely the following:

  - **An active response feature**: This is the feature where an IDS or IPS sends out a packet, such as TCP RST, in order to terminate a malicious connection.

  - **An inline mode of operation**: This is the case where a device or system operates in an inline mode, or as a bump in the wire. The following figure shows both offline and inline modes of operation. As you can see, the offline mode involves copying the packet and delivering it to the analysis module. In the case of inline operation, the system must analyze the packet and retransmit it to the network. There are many DAQ modules that support this, and each has its own way of implementing it.

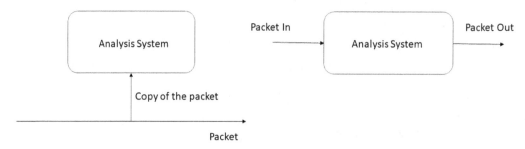

Figure 6.1 – The typical mode of operation for IDS (passive) and IPS (inline)

The need to transmit a packet from a system is a rare case in the case of IDS, whereas it is absolutely necessary in the case of IPS.

One main difference between IDS (passive) mode and IPS (inline) mode is how a system responds when its performance degrades. Since the IPS works in an inline fashion, performance degradation results in reduced throughput for all connections traversing a device, while in IDS mode, the result is packet drops and potential missed attacks. In the next section, let's discuss the performance of the DAQ layer.

The performance of the DAQ Layer

In addition to the functionality of packet I/O (delivery and transmission), there is another crucial requirement – namely, the performance aspect of this functionality. The packet data must be delivered most efficiently, taking as few CPU cycles as possible. This is important both in offline and inline scenarios.

Factors affecting packet capture performance

From an operational point of view, if the performance of a DAQ is at an undesirable level, it will result in degraded performance for an IDS and/or an IPS. The common factors that affect the performance of packet capture include the following:

- **Kernel–user space context switching**: Since programs such as Snort run in the user space, a system call is required to retrieve the packet from the kernel space to the user space. This becomes an overhead during a high traffic load. This is the case with the pcap DAQ. This is solved using a memory-mapped buffer (between the kernel and user space).

- **Data copying**: In the case of libpcap and similar DAQs, the packet data is copied several times – from the **Network Interface Card** (**NIC**) to the kernel space, from the kernel space to the user space, and so on. During a high traffic load, the number of memory copies increases and affects performance. This issue is resolved using an approach where the number of copies is reduced, such as directly copying the data from NIC to e user space, and zero-copy approaches.

Performance is one of the key factors for the research and development of new DAQ modules by open source communities and third parties.

The consequence of packet capture performance degradation

The packet capture mechanism is also part of the packet processing operation. When the traffic load increases, it needs to be processed at the same or faster rate than the traffic throughput rate. Handling temporary and short-lived bursts of traffic load can be helped by increasing the size of the packet buffer (memory) to store the unprocessed packets on the NIC card and/or kernel space.

The consequence of performance degradation in packet capture may include the following:

- **Packet drops**: This is the consequence in the case of IDS, which operate mostly in an offline and passive mode.

- **Packet latency (network throughput)**: In the case of IPS, the significant consequence is the degradation of network throughput due to packet drops and latency. Network protocols such as TCP react immediately when there is a packet drop anywhere along the path, whereas protocols such as UDP do not respond similarly. This feature of TCP protocol is called **Flow Control**. Thus, packets dropped due to IPS overload affects the TCP session throughput, which is not ideal.

In the next section, we will look at how Snort worked before the DAQ layer was introduced.

Packet capture in Snort

Let's take a brief look at how the packet capture functionality was implemented before DAQ was introduced (before Snort 2.9) and compare it to how this functionality is implemented in Snort 2.9.

Before DAQ

If we look at the Snort 2.8 code base, we can see that the packet capture functionality was ingrained within the Snort code, and it was not abstracted out as we have currently. This created unnecessary complexity and limitations in the Snort code. Let's take a look at the following code snippet from the `SnortMain()` function:

```
int SnortMain(int argc, char *argv[])
{
..
#ifdef GIDS
    if (InlineMode())
    {
#ifndef IPFW
        IpqLoop();
#else
        IpfwLoop();
#endif
    }
    else
    {
#endif /* GIDS */
        DEBUG_WRAP(DebugMessage(DEBUG_INIT, "Entering pcap loop\n"););
        InterfaceThread(NULL);
#ifdef GIDS
    }
..
```

In Snort and similar IDS systems, there is typically a *Packet Loop* method. This method implements a *loop* logic where the code reads a packet and performs the analysis in a loop. In the preceding snippet, we can see that there are different loop functions based on the mechanism for getting the packet – libpcap or libipq (IPTables), or libipfw **Internet Packet FireWall (IPFW)**.

In the preceding snippet, we can see that there are effectively three separate Packet Loop functions – namely, `IpqLoop`, `IpfwLoop`, and `InterfaceThread`. All these functions were defined within the Snort code base. (`InterfaceThread` is part of `snort.c`, and the other two functions are defined in `inline.c`.)

The `InterfaceThread` function handles the case where packets are read in a passive manner (offline as opposed to inline), and in the following snippet, we can see how it calls the libpcap methods (`pcap_loop` or `pcap_read`) to get packet data:

```
void *InterfaceThread(void *arg)
{
..

    /* Read all packets on the device.  Continue until cnt packets
read */
#ifdef USE_PCAP_LOOP
    pcap_ret = pcap_loop(pd, pv.pkt_cnt, (pcap_handler)
PcapProcessPacket, NULL);
#else
    while(1)
    {
..

        pcap_ret = pcap_dispatch(pd, pkts_to_read, (pcap_handler)
PcapProcessPacket, NULL);
```

In the case of inline mode, if IPQ is used, then the `IpqLoop()` function is called, which in turn calls `ipq_read()`. The `ipq_read()` function is provided by the libipq library, which interacts with IPTables to get the packet to the IDS or IPS. Alternatively, if IPFW is specified, then the `IpfwLoop()` method is called, which in turn calls socket APIs to retrieve the packet payload.

The code must be compiled for an inline operation (**Gateway IDS** or **GIDS**). If the code is compiled for an inline operation, then the functionality (inline versus offline) is decided by runtime mode. This is checked by a function called `InlineMode()`.

If we did not introduce the abstraction of DAQ, the code would get more and more complex for each additional packet capture library that Snort supports.

The DAQ module – introduced in Snort 2.9

The DAQ library and abstraction were added in Snort 2.9. In this section, let's take a look at how this interface was simplified by the use of DAQ. Let's look at the `PacketLoop` function in Snort 2.9:

```
void PacketLoop (void)
{
..

    while ( !exit_logged )
    {
        error = DAQ_Acquire(pkts_to_read, PacketCallback, NULL);
        if ( error )
        {
..
```

The `PacketLoop` function is much simpler and cleaner, since the entire details of specific packet capture mechanisms are abstracted out of Snort. These details are left for the DAQ library to handle. As far as Snort is concerned, the call to `DAQ_Acquire()` does the job of getting the packet and giving it to the packet processing function (`PacketCallback`).

In the next section, let's look at how the DAQ functionality is implemented in Snort 3.

The Snort 3 implementation of the DAQ layer

The DAQ layer abstraction separates out the logic of packet acquisition and the related functions into the DAQ library – libdaq. This library is available as a separate package from Snort. The libdaq library essentially consists of two parts – namely, the libdaq API and libdaq modules. All the DAQ library API functions can be found in the `api` directory (for example, `libdaq-3.0.5/api/`), whereas all the code for various DAQ modules can be found in the modules directory (for example, `libdaq-3.0.5/modules`). In the subsequent sections, we will examine these two DAQ library parts.

The DAQ library API

The DAQ Library API is a set of functions that can be called by Snort (or similar programs) without knowing the internal details of how the DAQ module is implemented. These APIs are grouped by their purpose:

- **Loading, unloading, and handling functions**: This set of DAQ API functions deals with the loading, unloading, and handling of the DAQ module. Functions to load/unload DAQ modules include `daq_load_static_modules`, `daq_load_dynamic_modules`, `daq_find_module`, and `daq_unload_modules`. The various DAQ library API functions in the preceding categories are listed in the `daq.h` header file (`api/daq.h`).

- **DAQ module functions**: These cover basic functions such as getting the name, version, and type of the DAQ module.

- **DAQ module configuration functions**: The DAQ API includes a set of functions for managing, setting, and retrieving configuration parameters for the DAQ module. The DAQ API includes a set of functions for managing, setting, and retrieving configuration parameters for the DAQ module. These DAQ module functions for managing config include functions such as `daq_module_config_new`, `daq_module_config_get_module`, `daq_module_config_set_mode`, `daq_module_config_set_variable`, and `daq_module_config_get_variable`.

- **DAQ configuration functions**: These are the set of functions to manage the configuration parameters for the DAQ from a user level. DAQ configuration functions include `daq_config_new`, `daq_config_set_input`, `daq_config_set_msg_pool_size`, `daq_config_set_snaplen`, and `daq_config_set_timeout`.

- **DAQ module instance functions**: Finally, there is a set of DAQ APIs related to the DAQ module instance. These include methods such as `daq_instance_instantiate`, `daq_instance_destroy`, `daq_instance_set_filter`, `daq_instance_start`, `daq_instance_inject`, and `daq_instance_msg_receive`.

In the next section, we will look at the various DAQ modules and their details.

DAQ modules

DAQ modules implement the underlying features and functionality of the packet capture from various data sources. In this section, we will look at some of the DAQ modules that come with Snort 3. We will also briefly discuss each of these DAQ modules. We will start with the legacy libpcap DAQ module.

libpcap

The pcap DAQ module depends on and builds on top of the libpcap library. The data source in the case of the pcap DAQ is network interfaces or network **packet capture (pcap)** files.

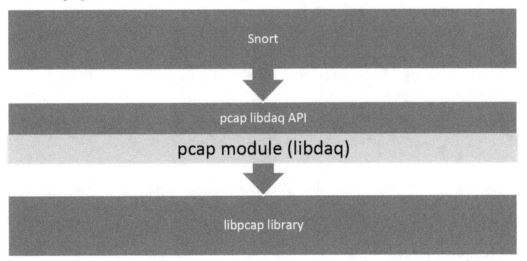

Figure 6.2 – Snort uses pcap module (libdaq) which in turn uses the libpcap library

The functions that are implemented by the pcap DAQ module are specified by the `DAQ_ModuleAPI_t` structure that is specified in the corresponding source file (`modules/pcap/daq_pcap.c`):

```
const DAQ_ModuleAPI_t pcap_daq_module_data =
{
    /* .api_version = */ DAQ_MODULE_API_VERSION,
    /* .api_size = */ sizeof(DAQ_ModuleAPI_t),
    /* .module_version = */ DAQ_PCAP_VERSION,
```

```
        /* .name = */ "pcap",
        /* .type = */ DAQ_TYPE_FILE_CAPABLE | DAQ_TYPE_INTF_CAPABLE | DAQ_
TYPE_MULTI_INSTANCE,
        /* .load = */ pcap_daq_module_load,
        /* .unload = */ pcap_daq_module_unload,
        /* .get_variable_descs = */ pcap_daq_get_variable_descs,
        /* .instantiate = */ pcap_daq_instantiate,
        /* .destroy = */ pcap_daq_destroy,
        /* .set_filter = */ pcap_daq_set_filter,
        /* .start = */ pcap_daq_start,
        /* .inject = */ pcap_daq_inject,
        /* .inject_relative = */ NULL,
        /* .interrupt = */ pcap_daq_interrupt,
        /* .stop = */ pcap_daq_stop,
        /* .ioctl = */ NULL,
        /* .get_stats = */ pcap_daq_get_stats,
        /* .reset_stats = */ pcap_daq_reset_stats,
        /* .get_snaplen = */ pcap_daq_get_snaplen,
        /* .get_capabilities = */ pcap_daq_get_capabilities,
        /* .get_datalink_type = */ pcap_daq_get_datalink_type,
        /* .config_load = */ NULL,
        /* .config_swap = */ NULL,
        /* .config_free = */ NULL,
        /* .msg_receive = */ pcap_daq_msg_receive,
        /* .msg_finalize = */ pcap_daq_msg_finalize,
        /* .get_msg_pool_info = */ pcap_daq_get_msg_pool_info,
    };
```

Note that the DAQ_ModuleAPI_t structure defined for the pcap DAQ module has several methods defined as NULL. Each DAQ module may not define all the methods but, rather, only the ones that apply to it.

netfilter queue (nfq)

The nfq DAQ module depends on and builds on top of the Linux netfilter packet filtering framework. Typically, this works in conjunction with IPTables (the Linux firewall) with a target of NFQUEUE. In this mechanism, the packet is queued (using the target NFQUEUE) to the user space for a verdict.

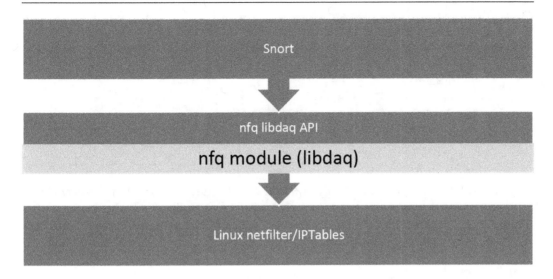

Figure 6.3 – Snort uses nfq module (libdaq), which in turn uses Linux netfilter/IPTables

The functions that are implemented by the nfq daq module are specified by the DAQ_ModuleAPI_t structure that is specified in the corresponding source file (modules/nfq/daq_nfq.c):

```
const DAQ_ModuleAPI_t nfq_daq_module_data =
{
    /* .api_version = */ DAQ_MODULE_API_VERSION,
    /* .api_size = */ sizeof(DAQ_ModuleAPI_t),
    /* .module_version = */ DAQ_NFQ_VERSION,
    /* .name = */ "nfq",
    /* .type = */ DAQ_TYPE_INTF_CAPABLE | DAQ_TYPE_INLINE_CAPABLE |
DAQ_TYPE_MULTI_INSTANCE | DAQ_TYPE_NO_UNPRIV,
    /* .load = */ nfq_daq_module_load,
    /* .unload = */ nfq_daq_module_unload,
    /* .get_variable_descs = */ nfq_daq_get_variable_descs,
    /* .instantiate = */ nfq_daq_instantiate,
    /* .destroy = */ nfq_daq_destroy,
    /* .set_filter = */ NULL,
    /* .start = */ nfq_daq_start,
    /* .inject = */ NULL,
    /* .inject_relative = */ NULL,
    /* .interrupt = */ nfq_daq_interrupt,
    /* .stop = */ nfq_daq_stop,
    /* .ioctl = */ NULL,
    /* .get_stats = */ nfq_daq_get_stats,
    /* .reset_stats = */ nfq_daq_reset_stats,
    /* .get_snaplen = */ nfq_daq_get_snaplen,
```

```
        /* .get_capabilities = */ nfq_daq_get_capabilities,
        /* .get_datalink_type = */ nfq_daq_get_datalink_type,
        /* .config_load = */ NULL,
        /* .config_swap = */ NULL,
        /* .config_free = */ NULL,
        /* .msg_receive = */ nfq_daq_msg_receive,
        /* .msg_finalize = */ nfq_daq_msg_finalize,
        /* .get_msg_pool_info = */ nfq_daq_get_msg_pool_info,
};
```

As in the pcap DAQ case, note that the DAQ_ModuleAPI_t structure defined for the nfq daq module also has several methods defined as NULL. In the case of nfq, specifically, we note that the inject method is defined as NULL. The inject feature injects a custom packet to the network (for example, a TCP RST packet) to close a malicious connection. This feature is not supported by the nfq DAQ.

Other DAQ modules

Besides the pcap and nfq DAQ modules, there are several more DAQ modules, such as the following:

- **afpacket**: The afpacket DAQ module enables direct access to raw packets from network devices. The afpacket DAQ uses Linux memory-mapped packet socket interfaces.

- **Divert**: The data source for Divert DAQ is BSD divert sockets. This is mainly for the BSD platform.

- **Netmap**: This is a DAQ module that is built on top of the netmap framework, which is a high-speed packet I/O framework. Linux and BSD systems support this DAQ module.

In addition to the preceding DAQ modules, there are third-party DAQ modules that are designed for specific purposes and special network cards. These are outside the scope of this book and will not be discussed here. In the next section, we will discuss how to configure the DAQ modules.

Configuring DAQ

The configuration options for the daq module can be listed using the help command available with Snort, as follows:

```
snort3 --help-module daq
```

The output for the preceding command is given here:

```
Configuration:
string daq.module_dirs[].path: directory path
string daq.inputs[].input: input source
int daq.snaplen = 1518: set snap length (same as -s) { 0:65535 }
int daq.batch_size = 64: set receive batch size (same as --daq-batch-
size) { 1: }
```

```
string daq.modules[].name: DAQ module name (required)
enum daq.modules[].mode = 'passive': DAQ module mode { 'passive' |
'inline' | 'read-file' }
string daq.modules[].variables[].variable: DAQ module variable
(foo[=bar])
```

These configuration parameters can be set within the lua configuration file.

In addition, there are the following command-line arguments for Snort that are relevant for DAQ:

`--daq <type>`	The option to select a packet acquisition module (the default is pcap) – for example, –daq nfq.
`--daq-batch-size`	This option sets the size of the batch received for processing.
`--daq-dir <dir>`	This option is used to specify the location of DAQ library files.
`--daq-list`	This option will list out the available DAQ choices.
`--daq-mode <mode>`	Use this option to select the DAQ module operating mode – for example, passive, inline, or read-file. This setting overrides the automatic selection.
`--daq-var <name=value>`	This option is used to set DAQ-related variables in a name=value format.

Table 6.1 – Command-line arguments for Snort related to DAQ

The –daq-list option lists out all the DAQs that are available for use. This list depends on the various DAQ modules included during build time:

```
snort --daq-list
Available DAQ modules:
afpacket(v7): live inline multi unpriv
 Variables:
  buffer_size_mb <arg> - Packet buffer space to allocate in megabytes
  debug - Enable debugging output to stdout
  fanout_type <arg> - Fanout loadbalancing method
  fanout_flag <arg> - Fanout loadbalancing option
  use_tx_ring - Use memory-mapped TX ring
nfq(v8): live inline multi
 Variables:
  debug - Enable debugging output to stdout
  fail_open - Allow the kernel to bypass the netfilter queue when it
is full
```

```
   queue_maxlen <arg> - Maximum queue length (default: 1024)
pcap(v4): readback live multi unpriv
 Variables:
  buffer_size <arg> - Packet buffer space to allocate in bytes
  no_promiscuous - Disables opening the interface in promiscuous mode
  no_immediate - Disables immediate mode for traffic capture (may
cause unbounded blocking)
  readback_timeout - Return timeout receive status in file readback
mode
```

In addition to the names of the various available DAQ modules, the –daq-list option also lists the variables that are relevant for each DAQ type.

As an example, let's see how we would run Snort using afpacket as the DAQ.

```
sudo snort -c ../../lua/snort.lua -R ~/snort-source-files/snort3-
community-rules/snort3-community.rules -i eth0 -A alert_fast -k none
--daq afpacket
```

If we want to modify one of the variables relevant to the DAQ, we can use the command-line option, --daq-var <name=value>, to do that. In the following example, we can see how the debug variable for the afpacket DAQ module is modified:

```
sudo ./snort -c ../../lua/snort.lua -R ~/snort-source-files/snort3-
community-rules/snort3-community.rules -i eth0 -A alert_fast -k none
--daq afpacket --daq-var debug=true
--------------------------------------------------
afpacket DAQ configured to passive.
Commencing packet processing
[eth0]
  TPacket Version: 1
  TPacket Header Length: 32
  MTU: 1500
  Reservation: 4
AFPacket Layout:
  Frame Size: 1600
  Frames:    83835
  Block Size: 131072 (Order 5)
  Blocks:    1035
  Wasted:    1523520
Created a ring of type 5 with total size of 135659520
++ [0] eth0
```

Among the listed DAQ modules, some support passive mode only, others support inline mode only, and a few support both modes.

Summary

Data acquisition is one of the primary and critical functionalities in the Snort architecture. This functionality also contributes directly to the performance of the entire system. We discussed the essential functionality of the DAQ layer and the module's performance aspects. We looked at how this functionality was implemented in Snort before the DAQ module was incorporated (in Snort 2.9), and we also looked briefly at the Snort 2.9 implementation of DAQ. Finally, we delved into DAQ functionality within Snort 3, and we discussed both the API side of DAQ and the various modules of DAQ that are currently supported.

In the next chapter, we will discuss the Codec module and its role in Snort IDS/IPS.

7
Packet Decoding

Packet decoding is the process of inspecting and interpreting the various protocol headers in a network packet. Every network packet consists of various encapsulation headers in addition to the data that it carries. When Snort analyzes an HTTP request packet, it performs decoding of all the protocol layers that encapsulate the HTTP request, starting from the outermost layer and working its way to the innermost layer – Ethernet, IPv4, and TCP. Each of these headers deals with various aspects of the communication – for example, the **Internet Protocol (IP)** header deals with aspects of sending the packet from one host (IP address) to another host (IP address), whereas the transport protocol header deals with ensuring reliable, consistent data transmission. In this chapter, we will study how Snort analyzes and decodes the various packet headers. We will also investigate how the packet decoding module is structured, the important data structures, and how the module ties to the rest of the system.

In this chapter, we're going to cover the following main topics:

- OSI layering and packet structure
- The role of packet decoding (Codecs)
- Packet decoding in Snort 3
- EthCodec – A layer 2 codec
- IPv4Codec – A layer 3 codec
- TcpCodec – A layer 4 codec
- Code structure and other codecs

OSI layering and packet structure

The network protocol stack (TCP/IP and similar protocol stacks) evolved to make it possible for various applications running on various computers to send and receive data from each other. Over time, each layer in the protocol stack was developed to take care of certain aspects of this data transfer. For example, one layer ensures the reliability and consistency of transferred data, while another layer makes sure that the data traverses from the source computer to the destination computer (addressing aspect). The TCP/IP protocol stack was created gradually by developing various protocol layers that served specific purposes.

The **Open Systems Interconnection** (**OSI**) layering model describes how a message is encapsulated with various headers – corresponding to the various (seven) layers – for communicating between two systems over the internet. The seven layers described in the OSI model include the physical layer, the data link layer, the network layer, the transport layer, the session layer, the presentation layer, and the application layer. The OSI model is a theoretical model.

The protocol suite that we encounter more frequently is the TCP/IP protocol suite, which has four layers. The layers in the OSI model can be mapped onto the TCP/IP protocol suite. The layers of the TCP/IP suite are as follows:

- Data link layer
- Network layer
- Transport layer
- Application layer

Let us look at an example of a packet to understand these four layers and headers; specifically, we will look at an HTTP GET request packet/frame:

```
>  Frame 4: 149 bytes on wire (1192 bits), 149 bytes captured (1192 bits)
>  Ethernet II, Src: 45:c6:aa:51:0c:98 (45:c6:aa:51:0c:98), Dst: 66:8f:bb:1d:9b:4a (66:8f:bb:1d:9b:4a)
>  Internet Protocol Version 4, Src: 192.168.140.190, Dst: 172.16.170.52
>  Transmission Control Protocol, Src Port: 32200, Dst Port: 80, Seq: 1, Ack: 1, Len: 95
>  Hypertext Transfer Protocol
```

```
0000  66 8f bb 1d 9b 4a 45 c6  aa 51 0c 98 08 00 45 00   f····JE· ·Q····E·
0010  00 87 00 01 00 00 40 06  d6 c4 c0 a8 8c be ac 10   ······@· ········
0020  aa 34 7d c8 00 50 00 00  00 0b 00 00 00 65 50 18   ·4}··P·· ·····eP·
0030  20 00 da 9d 00 00 47 45  54 20 2f 31 49 30 6a 42    ·····GE T /1I0jB
0040  2e 74 6f 72 72 65 6e 74  20 48 54 54 50 2f 31 2e   .torrent  HTTP/1.
0050  31 0d 0a 48 6f 73 74 3a  20 69 70 6c 6f 67 67 65   1··Host:  iplogge
0060  72 2e 6f 72 67 0d 0a 55  73 65 72 2d 41 67 65 6e   r.org··U ser-Agen
0070  74 3a 20 49 6d 70 6f 73  74 65 72 0d 0a 52 65 66   t: Impos ter··Ref
0080  65 72 65 72 3a 20 36 31  42 2d 41 42 38 2d 34 46   erer: 61 B-AB8-4F
0090  33 0d 0a 0d 0a                                     3····
```

Figure 7.1 – The data link layer (Ethernet layer header) for an HTTP GET request

Figure 7.1 shows the Wireshark view of an HTTP GET request. The top pane displays the various sections (e.g., Ethernet, Internet Protocol, etc.), and the bottom pane shows the packet bytes. Notice that the bytes highlighted in the bottom pane correspond to the selected section in the top pane. *Figure 7.1* highlights the Ethernet header which comprises 14 bytes.

Next, we will look at the IP header of the same packet:

```
>  Frame 4: 149 bytes on wire (1192 bits), 149 bytes captured (1192 bits)
>  Ethernet II, Src: 45:c6:aa:51:0c:98 (45:c6:aa:51:0c:98), Dst: 66:8f:bb:1d:9b:4a (66:8f:bb:1d:9b:4a)
>  Internet Protocol Version 4, Src: 192.168.140.190, Dst: 172.16.170.52
>  Transmission Control Protocol, Src Port: 32200, Dst Port: 80, Seq: 1, Ack: 1, Len: 95
>  Hypertext Transfer Protocol

0000   66 8f bb 1d 9b 4a 45 c6   aa 51 0c 98 08 00 45 00   f····JE· ·Q····E·
0010   00 87 00 01 00 00 40 06   d6 c4 c0 a8 8c be ac 10   ······@· ········
0020   aa 34 7d c8 00 50 00 00   00 0b 00 00 00 65 50 18   ·4}··P·· ·····eP·
0030   20 00 da 9d 00 00 47 45   54 20 2f 31 49 30 6a 42   ·····GE T /1I0jB
0040   2e 74 6f 72 72 65 6e 74   20 48 54 54 50 2f 31 2e   .torrent  HTTP/1.
0050   31 0d 0a 48 6f 73 74 3a   20 69 70 6c 6f 67 67 65   1··Host:  iplogge
0060   72 2e 6f 72 67 0d 0a 55   73 65 72 2d 41 67 65 6e   r.org··U ser-Agen
0070   74 3a 20 49 6d 70 6f 73   74 65 72 0d 0a 52 65 66   t: Impos ter··Ref
0080   65 72 65 72 3a 20 36 31   42 2d 41 42 38 2d 34 46   erer: 61 B-AB8-4F
0090   33 0d 0a 0d 0a                                      3····
```

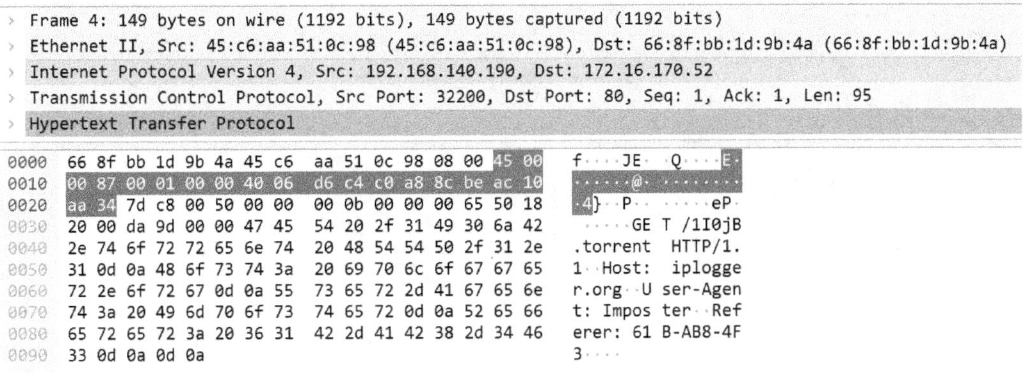

Figure 7.2 – The network layer (Internet Protocol header) for the HTTP GET request

The IP header is 20 bytes in this case. The highlighted portion of the bottom pane shows the bytes in the IP header. Next, we will look at the TCP header of the same packet:

```
>  Frame 4: 149 bytes on wire (1192 bits), 149 bytes captured (1192 bits)
>  Ethernet II, Src: 45:c6:aa:51:0c:98 (45:c6:aa:51:0c:98), Dst: 66:8f:bb:1d:9b:4a (66:8f:bb:1d:9b:4a)
>  Internet Protocol Version 4, Src: 192.168.140.190, Dst: 172.16.170.52
>  Transmission Control Protocol, Src Port: 32200, Dst Port: 80, Seq: 1, Ack: 1, Len: 95
>  Hypertext Transfer Protocol

0000   66 8f bb 1d 9b 4a 45 c6   aa 51 0c 98 08 00 45 00   f····JE· ·Q····E·
0010   00 87 00 01 00 00 40 06   d6 c4 c0 a8 8c be ac 10   ······@· ········
0020   aa 34 7d c8 00 50 00 00   00 0b 00 00 00 65 50 18   ·4}··P·· ·····eP·
0030   20 00 da 9d 00 00 47 45   54 20 2f 31 49 30 6a 42   ·····GE T /1I0jB
0040   2e 74 6f 72 72 65 6e 74   20 48 54 54 50 2f 31 2e   .torrent  HTTP/1.
0050   31 0d 0a 48 6f 73 74 3a   20 69 70 6c 6f 67 67 65   1··Host:  iplogge
0060   72 2e 6f 72 67 0d 0a 55   73 65 72 2d 41 67 65 6e   r.org··U ser-Agen
0070   74 3a 20 49 6d 70 6f 73   74 65 72 0d 0a 52 65 66   t: Impos ter··Ref
0080   65 72 65 72 3a 20 36 31   42 2d 41 42 38 2d 34 46   erer: 61 B-AB8-4F
0090   33 0d 0a 0d 0a                                      3····
```

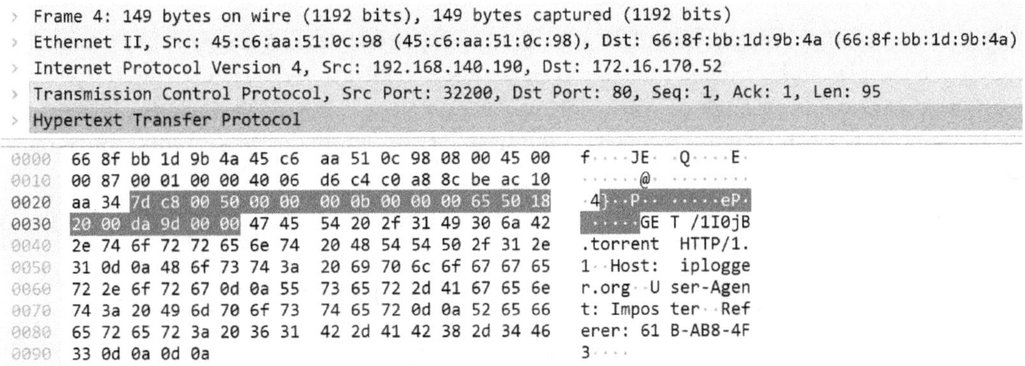

Figure 7.3 – The transport layer (Transmission Control Protocol header) for the HTTP GET request

The TCP header is 20 bytes in this case. The highlighted portion of the bottom pane shows the bytes in the TCP header. Finally, we will look at the application data portion of the same packet – which is the HTTP request:

```
> Frame 4: 149 bytes on wire (1192 bits), 149 bytes captured (1192 bits)
> Ethernet II, Src: 45:c6:aa:51:0c:98 (45:c6:aa:51:0c:98), Dst: 66:8f:bb:1d:9b:4a (66:8f:bb:1d:9b:4a)
> Internet Protocol Version 4, Src: 192.168.140.190, Dst: 172.16.170.52
> Transmission Control Protocol, Src Port: 32200, Dst Port: 80, Seq: 1, Ack: 1, Len: 95
> Hypertext Transfer Protocol
```

```
0000   66 8f bb 1d 9b 4a 45 c6  aa 51 0c 98 08 00 45 00   f····JE· ·Q····E·
0010   00 87 00 01 00 00 40 06  d6 c4 c0 a8 8c be ac 10   ······@· ········
0020   aa 34 7d c8 00 50 00 00  00 0b 00 00 00 65 50 18   ·4}··P·· ·····eP·
0030   20 00 da 9d 00 00 47 45  54 20 2f 31 49 30 6a 42    ·····GE T /1I0jB
0040   2e 74 6f 72 72 65 6e 74  20 48 54 54 50 2f 31 2e   .torrent  HTTP/1.
0050   31 0d 0a 48 6f 73 74 3a  20 69 70 6c 6f 67 67 65   1··Host:  iplogge
0060   72 2e 6f 72 67 0d 0a 55  73 65 72 2d 41 67 65 6e   r.org··U ser-Agen
0070   74 3a 20 49 6d 70 6f 73  74 65 72 0d 0a 52 65 66   t: Impos ter··Ref
0080   65 72 65 72 3a 20 36 31  42 2d 41 42 38 2d 34 46   erer: 61 B-AB8-4F
0090   33 0d 0a 0d 0a                                     3····
```

Figure 7.4 – The application layer (HTTP GET request)

It is the job of the Codecs module to parse out and interpret the protocol headers (except the application layer, which will be done by subsequent modules, for example, by inspector modules).

Data encapsulation and decapsulation

The process of adding a header to the data is called **encapsulation**, and the reverse process of removing the header and retrieving the data is called **decapsulation**. As the data being sent from one application traverses through one protocol layer at a time, each protocol layer adds a header (a layer of encapsulation) and, finally, the packet with several headers and the data is sent through the network to the destination.

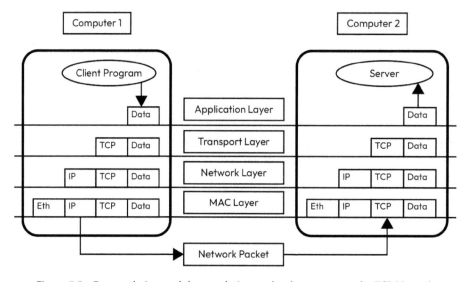

Figure 7.5 – Encapsulation and decapsulation as the data traverses the TCP/IP stack

When this packet is received at the destination, the packet goes through the process of decapsulation – each layer interprets and removes the respective header and passes the data to the upper protocol layer. Finally, the data reaches the receiving application having stripped off every protocol header. The Codecs in Snort must do the decapsulation of the packet as the TCP/IP stack of the receiving host. We will discuss this in the next section.

The role of packet decoding (Codecs)

As mentioned earlier, packet decoding is the process of inspecting and interpreting the various protocol headers in a network packet. The process is very similar to the decapsulation and decoding of network packets at the destination host. The decoding process also includes the necessary validation of the header fields; for example, the checksum values will be validated. Any abnormalities in the protocol headers are detected during this phase. These may be due to benign or malicious reasons. The decoding phase may generate alerts for such abnormalities if the NIDS is configured accordingly.

Codecs do not perform complex tasks such as session management or IP fragmentation reassembly. Rather, they decode the respective protocol headers (and populate the corresponding internal data structures) so as to enable the subsequent modules (such as inspectors) to perform their role.

In addition to performing the decapsulation process for every packet, the NIDS also must interpret each protocol layer header and look for any malicious activity that may be present in the various protocol layer headers. For example, various attacks can happen just by sending packets with invalid protocol headers. It is the role of the NIDS to interpret the protocol headers and detect any such malicious activity.

Packet decoding in Snort 3

In the previous sections, we have seen how the encapsulation and decapsulation process works, and the role of packet decoding from an IDS/IPS point of view. In this section, we will look at how this is done in Snort 3; the module that does this functionality in Snort 3 is called the **Codecs module**.

Once the packet has been acquired by the DAQ layer, the packet data is processed and analyzed by the appropriate codecs. In the case where the data link layer is Ethernet, the first codec that analyzes the packet will be EthCodec. EthCodec parses the packet data based on the `Ethernet` header. The `Ethernet` header specifies the type of network protocol using the `type` field (2 bytes). The hex value of `0x0800` indicates that the network protocol is IP. If this is the case, the next codec that will handle the packet data will be the IPv4 codec. The transport layer protocol is specified by the `protocol` field (2 bytes) within the IPv4 header. The common values for transport layer protocols are TCP (protocol value of `6`), and UDP (protocol value of `17`). If the transport layer protocol is TCP, the next Codec to handle the packet data would be TcpCodec. As it can be noted, the specific codecs that will be invoked to decode the packet headers depend on the type of protocol layers that are present in the packet data. The operation that is performed by the Codec module is very similar to the decapsulation process within the typical network stack.

While each codec decodes the packet data, if it detects any validation errors, header errors, or any indication of malicious behavior, a codec event (alert) is generated. We will look at a few specific cases of codec events when we look at a few specific codecs next.

In the next sections, we will look closely at some of the common codecs – namely, EthCodec, IPv4Codec, and TCPCodec – as examples of layer 2, layer 3, and layer 4 codec functionality.

EthCodec – a layer 2 codec

EthCodec is a data link layer codec and is one of the top-level codecs. A top-level codec is one of the codecs that starts the packet decoding functionality; the data link layer (in this case, the Ethernet layer) is the outermost header that needs to be decoded first before other layers can be decoded.

There are not many validations done at this layer. The codec checks whether the total length of the packet data handed over is less than the `Ethernet` header length, which is 14 bytes. If this is indeed the case, a codec event is generated, namely, a *truncated Ethernet header*.

In normal cases, the `Ethernet` header will not be truncated (the packet data length will be more than 14 bytes). After this validation is checked, the codec lays the `Ethernet` header structure over the packet data and checks the `type` field. The `Ethernet` header structure is defined in the `src/protocols/eth.h` header file and is as follows:

```
struct EtherHdr
{
    uint8_t ether_dst[6];
    uint8_t ether_src[6];
    uint16_t ether_type;
    /* return data in byte order */
    inline ProtocolId ethertype() const
    { return (ProtocolId)ntohs(ether_type); }
    /* return data in network order */
    inline uint16_t raw_ethertype() const
    { return ether_type; }
};
```

If we look back at the example from *Figure 7.1*, the `Ethernet` header bytes were as follows (in hex): `66 8f bb 1d 9b 4a 45 c6 aa 51 0c 98 08 00`. Let us lay the preceding Ethernet structure over these data bytes:

Ethernet Header Length (14 bytes)		
ether_dst (6 bytes)	ether_src (6 bytes)	ether_type (2 bytes)
66 8f bb 1d 9b 4a	45 c6 aa 51 0c 98	08 00

Table 7.1 – Example shows the mapping of the Ethernet header to the raw bytes sequence

Once you lay the Ethernet header structure over the packet bytes, we see the individual field values. In this case, the type is 0x 08 00, which maps to the IPv4 protocol as the network protocol. We will look at IPv4Codec in the next section.

IPv4Codec – a layer 3 codec

IPv4Codec is invoked when the packet uses **IP version 4 (IPv4)** as the network protocol. IP is more detailed and involved than the Ethernet protocol, and hence the codec also does more validations and checks on the IP header.

The IP header structure is defined in the src/protocols/ipv4.h header file, and the structure is as follows:

```
struct IP4Hdr
{
    uint8_t ip_verhl;        /* version & header length */
    uint8_t ip_tos;          /* type of service */
    uint16_t ip_len;         /* datagram length */
    uint16_t ip_id;          /* identification  */
    uint16_t ip_off;         /* fragment offset */
    uint8_t ip_ttl;          /* time to live field */
    IpProtocol ip_proto;      /* datagram protocol */
    uint16_t ip_csum;        /* checksum */
    uint32_t ip_src;  /* source IP */
    uint32_t ip_dst;  /* dest IP */
<snip>
```

Only the relevant portion of the IP4Hdr structure is shown – that which is used to lay over the packet data bytes and access the various IP header fields. You may check out the entire structure in src/protocols/ipv4.h.

The IP header is a variable-length header. The length of the IP header can be found in the ip_verhl field. This field is one byte and the first four bits represent the IP protocol version, which, in IPv4, is always 4. The next four bits represent the IP header length. Since it is 4 bits, the value can go from 0 to 15, and the actual IP header length is 4 times this value.

There is a set of validations and checks that the codec performs on the IP header. The set of events (alerts) that are generated by the codec is given in the following table:

Ipv4 Codec Events
Not IPv4 datagram
IPv4 header length < minimum
IPv4 datagram length < header field
IPv4 options found with bad lengths
Truncated IPv4 options
IPv4 datagram length > captured length
IPv4 packet with zero TTL
IPv4 packet with bad frag bits (both MF and DF set)
IPv4 packet frag offset + length exceeds maximum
IPv4 packet from "current net" source address
IPv4 packet to "current net" dest address
IPv4 packet from the multicast source address
IPv4 packet from the reserved source address
IPv4 packet to reserved dest address
IPv4 packet from the broadcast source address
IPv4 packet to broadcast dest address
IPv4 packet below TTL limit
IPv4 packet both DF and offset set
IPv4 reserved bit set
IPv4 option set
Truncated IPv4 header
Bad checksum

Table 7.2 – The list of events or alerts raised by the IPv4 codec

The abbreviation MF in *Table 7.2* stands for "More Fragments", DF stands for "Don't Fragment" and TTL stands for "Time To Live". Each of these are fields in the IP header.

If we look back at the example from *Figure 7.1*, the IPv4 layer header bytes were as follows (in hex) – `45 00 00 87 00 01 00 00 40 06 d6 c4 c0 a8 8c be ac 10 aa 34`. Let us lay the preceding IP header structure over these data bytes:

45	00	00	87
ip_verhl	ip_tos	ip_len	
00	01	00	00
ip_id		offset	
40	06	d6	c4
ttl	proto	ip_csum	
c0	a8	8c	be
ip_srcc			
ac	10	aa	34
ip_dst			

Figure 7.6 – Laying the IP header structure over the relevant packet data bytes

As you lay the IP header structure over the packet bytes, we see the individual field values. The next layer (transport layer in this case) will be indicated by the protocol field (marked as `proto`). The protocol value here is 6, which is **Transmission Control Protocol (TCP)**. The transport layer will be decoded in turn by the appropriate Codec; in this case, it will be TcpCodec.

In the next section, we will look briefly at layer 4 decoding in Snort.

TcpCodec – a layer 4 codec

The layer 4 in the OSI model is also known as the transport layer. This layer deals with the establishment and maintenance of connections, flow control, and ensuring reliable communication. The layer 4 header will be found encapsulated within the Ipv4 (layer 3) datagram.

As we discussed in the previous section (in *Figure 7.6*), `proto` within the Ipv4 header denotes the type of layer 4 protocol. There are several protocols that fall in this layer. Some of the common layer 4 protocols include TCP (protocol value `4`), UDP (protocol value `17`), and ICMP (protocol value `1`). The following table shows a subset of the IP values and the corresponding layer 4 protocol:

IP proto value	Layer 4 protocol
1	Internet Control Message Protocol (ICMP)
2	Internet Group Management Protocol (IGMP)
3	Gateway to Gateway Protocol (GGP)
4	Ipv4 encapsulation
5	ST (Stream Protocol)
6	TCP (Transport Control Protocol)
8	EGP (Exterior Gateway Protocol)
9	IGP (Any private interior gateway protocol)
17	UDP (User Datagram Protocol)
47	GRE (Generic Routing Encapsulation Protocol)
50	ESP (Encapsulating Security Protocol)

Table 7.3 – A subset of IP protocol values and the corresponding layer 4 protocols

Snort has implemented codecs for the most common layer 4 protocols – namely, TCP, UDP, and ICMP. In addition, Snort has codecs for a handful of other layer 4 protocols as well. The layer 4 codecs supported by Snort 3 include the following: ICMP codec, IGMP codec, TCP codec, UDP codec, ESP codec, and GRE codec. As an example, we will discuss the TCP codec briefly in this section.

TcpCodec is invoked when the packet uses TCP as the layer 4 protocol. TCP is a connection-oriented, ordered, and reliable protocol for end-to-end data transmission. TCP establishes an end-to-end connection and uses sequence numbers to ensure successful data delivery. TCP payload is encapsulated within an Ipv4 datagram. After the Ipv4 codec has decoded the network header, TcpCodec takes over to decode the TCP header.

As part of the decoding process by TcpCodec, it fills the TCP header structure that is part of the packet structure. The TCP header struct is defined as follows:

```
struct TCPHdr
{
    uint16_t th_sport;      /* source port */
    uint16_t th_dport;      /* destination port */
    uint32_t th_seq;        /* sequence number */
    uint32_t th_ack;        /* acknowledgment number */
    uint8_t th_offx2;       /* offset and reserved */
    uint8_t th_flags;
```

```
uint16_t th_win;        /* window */
uint16_t th_sum;        /* checksum */
uint16_t th_urp;        /* urgent pointer */
```

Only the relevant portion of the `TCPHdr` structure is shown – that which is used to lay over the packet data bytes and access the various TCP header fields. You may check out the entire structure in `src/protocols/tcp.h`.

Like the IP header, the TCP header is also a variable-length header. The length of the IP header can be found in the `th_offx2` field. This field is one byte, and the first four bits represent the IP header length. Since it is 4 bits, the value can go from 0 to 15, and the actual TCP header length is 4 times this value.

Let us continue to look at the example from *Figure 7.1*; the TCP layer header bytes were as follows (in hex): 7d c8 00 50 00 00 00 0b 00 00 00 65 50 18 20 00 da 9d 00 00. In order to decode the TCP header, let us lay the preceding TCP header structure over these data bytes:

7d	c8	00	50
	th_sport		**th_dport**
00	00	00	0b
		th_seq	
00	00	00	65
		th_ack	
50	18	20	00
th_offx2	**th_flags**		**th_win**
da	9d	00	00
	th_sum		**th_urp**

Figure 7.7 – Laying the TCP header structure over the relevant packet data bytes

As you lay the TCP header structure over the packet bytes, we see the individual field values. For example, the destination port is 80 (`th_dport` is decoded as hex; 0x50 equates to 50 in decimal). The next layer (application layer in this case) will be indicated by the destination port field. Snort does not decode application layer data; when appropriate the application layer data is decoded by an appropriate inspector. For example, in this case, the application protocol is HTTP and will be handled by an HTTP inspector.

In the next section, we will look briefly at the various files and directory structures for the codec-related functionality.

Code structure and other codecs

Let us take a quick look at how the source files related to codecs are organized within the Snort 3 source code:

```
ls src/codecs
CMakeLists.txt  codec_api.cc  codec_api.h  codec_module.cc  codec_
module.h  dev_notes.txt  ip  link  misc  root
```

The source code relevant to codecs is organized under the directory named `codecs`. We see four directories: `ip`, `link`, `misc`, and `root`. Let us list the files under each of these directories as well:

```
ls src/codecs/root
cd_eth.cc  cd_raw.cc  CMakeLists.txt  dev_notes.txt
```

The `root` directory includes the common top-level codecs, `ethernet` and `raw`. These two are the common codecs that serve as the first-level decoders. Next, let us look at the `ip` directory:

```
ls src/codecs/ip
cd_auth.cc        cd_esp.cc    cd_hop_opts.cc  cd_igmp.cc  cd_mobility.
cc  cd_routing.cc  checksum.h
cd_bad_proto.cc  cd_frag.cc  cd_icmp4.cc     cd_ipv4.cc  cd_no_next.
cc    cd_tcp.cc      CMakeLists.txt
cd_dst_opts.cc    cd_gre.cc   cd_icmp6.cc     cd_ipv6.cc  cd_pgm.
cc        cd_udp.cc      dev_notes.txt
```

The `ip` directory comprises all the decoders that comprise the IP suite. These include protocols at various layers including the network and transport layers. The `cd_ipv4.cc`, `cd_frag.cc`, and `cd_ipv6.cc` source files deal with the network layer (OSI layer 3), whereas the `cd_tcp.cc` and `cd_udp.cc` source files deal with the transport layer (OSI layer 4) codec functionality.

Next, let us peek into the `link` directory:

```
ls src/codecs/link
cd_arp.cc                cd_erspan2.cc  cd_fabricpath.cc  cd_ppp_encap.
cc  cd_trans_bridge.cc  CMakeLists.txt
cd_ciscometadata.cc  cd_erspan3.cc  cd_mpls.cc        cd_pppoe.
cc        cd_vlan.cc         dev_notes.txt
```

The `link` directory has a set of codecs that are data link layer codecs. Finally, we will look at the `misc` directory. The `misc` directory includes a set of codecs that are general; for example, some tunneling protocol codecs are included here (`cd_gtp`):

```
ls src/codecs/misc
cd_default.cc  cd_gtp.cc        cd_icmp6_ip.cc  cd_teredo.cc  cd_vxlan.
cc       dev_notes.txt
cd_geneve.cc   cd_icmp4_ip.cc  cd_llc.cc        cd_user.
cc    CMakeLists.txt
```

In this section, we looked at how the source code for the codec module is structured. It is highly encouraged that you download the source code and get familiar with it.

Summary

This chapter introduced you to the codec layer or module of Snort 3. We saw the role of codecs, and also discussed the encapsulation and decapsulation processes that happen in every network stack and also in IDS/IPS. We used an example packet and discussed how the relevant codecs decoded the packet data in Snort 3. We looked at the layer 2 and layer 3 codec functionality in Snort 3. Finally, we looked at how the code is laid out and structured within Snort 3 code. In the next chapter, we will look into the TCP state tracking and inspectors.

8

Inspectors

The Snort 3 system performs in-depth analysis for a wide range of network protocols. It does traffic analysis on **Protocol Data Units (PDUs)** rather than packets. This protocol analysis logic is implemented as pluggable modules called **inspectors**.

Inspectors, as the backbone of Snort 3, play a pivotal role in its functioning. From a functionality standpoint, inspectors can be seen like the preprocessors in Snort 2. In other words, inspectors may be considered the successor of the preprocessor.

Snort 3 has a modular architecture, and each inspector is implemented as a plugin. Before we delve into the various modules implemented as inspectors, we should discuss inspectors in general. In this chapter, we're going to cover the following main topics:

- The role of inspectors
- Types of inspectors
- Snort 3 inspectors

The role of inspectors

Ideally, the Snort system should analyze the network traffic as an end host or server would do in order to detect malicious or otherwise interesting activity.

For an end host, traffic data is first processed by the TCP/IP layers (layers 2–4) before reaching specific applications. There is a large number and a wide variety of applications, such as the following:

- Web clients (browsers) such as Mozilla Firefox and web servers such as Apache, which use the HTTP/HTTPS protocol.
- Mail clients such as Outlook and mail servers such as Nginx, which use protocols such as SMTP, IMAP, POP, and so on.

Similarly, in the case of Snort 3, the decoders perform the TCP/IP layer analysis (layer 2 to layer 4 analysis), and then the inspectors do the application-specific analysis. These modules implement the analysis logic to mimic the essential analysis done by the various application client and server programs.

Architecturally, inspectors are implemented as configurable plugin modules designed to perform very specific tasks. Inspectors are versatile and the tasks are very varied, such as state tracking, data reassembly (IP fragmentation, TCP segmentation), port scan detection, traffic normalization, file analysis, and several application-level protocol analyses (HTTP, DCE RPC, SMTP, etc.).

Inspectors are implemented as C++ classes. During Snort initialization time, each of these inspectors is instantiated. Certain inspectors can only have a single instance; these are called **singletons**. Alternatively, other inspectors that can have several instances created are called **multiton** inspectors. The inspectors are enabled/disabled and configured at the policy level.

In previous Snort versions, the system design was such that the enabled preprocessors were maintained in a list, and all packets would go through all the preprocessors. When the traffic was not applicable to a preprocessor, the code would just skip any analysis and return.

In Snort 3, there are two special-purpose modules, namely, the **binder** inspector and the **wizard** inspector. Every flow is inspected by the binder inspector, and it is assigned, with the help of a wizard inspector, to the appropriate stream and service inspectors for further analysis.

Types of inspectors

Based on the type of analysis, the inspectors are grouped together into three categories, which are discussed in the following subsections.

Network inspectors

Network inspectors deal with the analysis and processing of network and transport layer protocol. These are a set of inspector modules that operate on the network layer of the processed traffic. From a functionality point of view, these network inspectors do a variety of functions, such as managing sessions, monitoring performance, binding incoming traffic to specific configurations, and so on. Although these are varied in functionality, they are grouped as network inspectors since they deal primarily with network layer analysis.

Service inspectors

Service inspectors deal with the analysis of various application-level protocols. These modules analyze and decode the application-level protocols (such as HTTP, SMTP, etc). These modules detect any protocol-related abnormalities and can generate specific alerts for these abnormalities if configured to do so. Service inspectors perform in-depth analysis and dissect the various protocol messages so as to enable more accurate IDS/IPS rule matching.

For example, the HTTP inspector module dissects the HTTP request into the method, URI, version, HTTP headers, and so on. This allows the rules to specifically match against these individual components.

Stream inspectors

Stream inspectors track the bi-directional flow, do state tracking, and also do reassembly as needed for the TCP, UDP, and ICMP protocols. These inspectors mainly deal with session tracking for the IP, TCP, UDP, and ICMP protocols. Of these, TCP is the only stateful protocol; the other protocols (IP, UDP, and ICMP) are stateless. State tracking, stream reassembly, and normalizer functionality are part of the TCP stream module.

Next, let us look at the different categories and the various inspectors that belong under them.

Snort 3 inspectors

In this section, we will look at the various specific inspectors that are grouped under the three categories. Let's use the --show-plugins command-line option of Snort to list the available inspectors:

```
./build/src/snort --show-plugins
```

The above command prints out all the available plugins grouped by types such as Codecs, Inspectors, Search Engine, etc. We find the list of inspectors under the section called *Inspectors* as shown in the following image:

```
Inspectors
                appid(v0)              arp_spoof(v0)            back_orifice(v0)
               binder(v0)                    cip(v0)          dce_http_proxy(v0)
      dce_http_server(v0)                dce_smb(v0)                 dce_tcp(v0)
              dce_udp(v0)                   dnp3(v0)                     dns(v0)
              file_id(v0)               file_log(v0)              ftp_client(v0)
             ftp_data(v0)             ftp_server(v0)             gtp_inspect(v0)
        http2_inspect(v0)           http_inspect(v0)                  iec104(v0)
                 imap(v0)                    mms(v0)                  modbus(v0)
              netflow(v0)             normalizer(v0)          packet_capture(v0)
         perf_monitor(v0)                    pop(v0)               port_scan(v0)
           reputation(v0)                    rna(v0)              rpc_decode(v0)
           s7commplus(v0)                    sip(v0)                    smtp(v0)
             so_proxy(v0)                    ssh(v0)                     ssl(v0)
               stream(v0)            stream_file(v0)              stream_icmp(v0)
            stream_ip(v0)             stream_tcp(v0)              stream_udp(v0)
          stream_user(v0)                 telnet(v0)                  wizard(v0)
```

Figure 8.1 – The list of inspectors (as printed using the --show-plugins option)

Another useful command that lists all the inspectors with some details about what the inspector does is given next (the entire output of the command is not shown here for brevity):

```
./build/src/snort --help-plugins | grep inspector
inspector::appid: application and service identification
inspector::arp_spoof: detect ARP attacks and anomalies
inspector::back_orifice: back orifice detection
inspector::binder: configure processing based on CIDRs, ports,
services, etc.
inspector::cip: cip inspection
inspector::dce_http_proxy: dce over http inspection - client to/from
proxy
inspector::dce_http_server: dce over http inspection - proxy to/from
server
```

In order to get information on any of the listed inspectors, we can use the --help-module command (shown here with the dns module as an example):

```
./build/src/snort --help-module dns
dns
Help: dns inspection
Type: inspector (service)
Usage: inspect
Instance Type: multiton
Configuration:
bool dns.publish_response = false: parse and publish dns responses
Rules:
131:1 (dns) obsolete DNS RR types
131:2 (dns) experimental DNS RR types
131:3 (dns) DNS client rdata txt overflow
Peg counts:
dns.packets: total packets processed (sum)
dns.requests: total dns requests (sum)
dns.responses: total dns responses (sum)
dns.concurrent_sessions: total concurrent dns sessions (now)
dns.max_concurrent_sessions: maximum concurrent dns sessions (max)
```

As we mentioned, these inspectors are categorized as network, service, and stream inspectors. We will start with the network inspectors in the next section.

Network inspectors

The inspectors that are grouped as network inspectors primarily deal with the analysis and processing of the network and transport layer protocols. The network inspectors include the following:

- **Binder**: This is a special inspector that maps a flow to the appropriate config (policy selection).

- **arp_spoof**: The **Address Resolution Protocol** (**ARP**) deals with the task of mapping an IP address to the Ethernet address, especially in the context of a local network (same broadcast domain). Various attacks happen at the ARP level, for example, ARP spoofing. This inspector analyzes the ARP, checks for consistency, and detects spoofing attempts and ARP cache inconsistencies.

- **port_scan**: This inspector detects various scanning attempts over various protocols including TCP, UDP, and ICMP.

- **Normalizer**: This inspector is relevant in an inline mode of operation. This module normalizes the network traffic in order to remove any ambiguity and thereby mitigate evasion attacks. A typical example would be a TCP segment overlap scenario; in this case, the normalizer would modify the packets such that there is no overlapping TCP data.

- **perf_monitor**: This inspector monitors for and checks against various performance criteria.

- **packet_capture**: This is a tool for dumping the wire packets that Snort receives.

Next, let us look at the various service inspectors.

Service inspectors

The service inspectors are the group of inspectors that deal with protocols that are above layer 4. The service inspectors are varied and cover several types of protocols, such as the following:

- **Mail-related**: A handful of inspectors cover mail-related protocols, namely, the IMAP, POP, and SMTP inspectors:

  - **IMAP inspector**: The IMAP inspector decodes and analyzes the **Internet Message Application Protocol** (**IMAP**), which is used by email clients to retrieve email messages from a remote IMAP server. IMAP uses TCP port 143 and TCP port 993 (encrypted).

  - **POP inspector**: The POP inspector decodes and analyzes the **Post Office Protocol version 3** (**POP3**), which email clients use to retrieve email messages from a remote server.

  - **SMTP inspector**: **Simple Mail Transfer Protocol** (**SMTP**) is an internet standard communication protocol for electronic mail transmission. It is one of the common protocols associated with email transmission. SMTP usually operates over TCP port 25. The SMTP inspector decodes and analyzes the SMTP protocol.

- **Web-related**: The HTTP inspectors cover the HTTP analysis, and the HTTP2 inspector covers the HTTP2 version:

  - **HTTP Inspect inspector**: This is a revamp of the HTTP Inspect module in Snort 2. The HTTP Inspect inspector analyzes HTTP and enables the detection of attacks over web traffic.

- **SCADA-related**: **Supervisory Control and Data Acquisition (SCADA)** systems are critical, and these systems use a set of special protocols. A handful of inspectors cover these protocols, namely, S7CommPlus, DNP3, IEC 104, Modbus, and CIP.

  - **CIP inspector**: The **Common Industrial Protocol (CIP)** is a protocol used for industrial automation applications and is often used in SCADA networks. The CIP inspector analyzes the CIP messages to enable the detection of attacks.

  - **DNP3 inspector**: The DNP3 inspector decodes and analyzes the DNP3 protocol, one of the SCADA-related protocols used with industrial and power automation systems.

  - **IEC 104 inspector**: This module analyzes IEC 104, yet another SCADA protocol that is used with power systems.

  - **Modbus inspector**: Modbus is a data communication protocol originally used with **Programmable Logic Controllers (PLCs)**, and nowadays, often used for communication with industrial electronic devices. The Modbus inspector decodes and analyzes the Modbus protocol traffic. Modbus also belongs to the SCADA suite of protocols.

  - **S7CommPlus inspector**: S7CommPlus is a proprietary protocol developed by Siemens to use with PLCs. S7CommPlus is also a SCADA protocol. This inspector inspects and decodes this custom protocol.

- **DNS-related**: The DNS inspector covers the analysis of the DNS protocol over both UDP and TCP.

- **NetFlow**: In addition, there is the NetFlow service inspector, which monitors NetFlow traffic between a NetFlow collector and a NetFlow exporter. This helps Snort 3 get information on new and existing flows and the associated services.

- **SSL**: The SSL inspector performs the analysis for the SSL and TLS protocols used in conjunction with the HTTPS protocol.

- **Miscellaneous**: In addition, there are a few more inspectors, such as `rpc_decode` for remote procedure calls, `ftp_telnet` for FTP and Telnet protocols, `dce_rpc`, `mms`, and `back_orifice`:

  - **DCE SMB inspector**: **Distributed Computing Environment/Remote Procedure Calls (DCE/RPC)** is an RPC system that can work over various layers including SMB, TCP, UDP, and so on. The DCE SMB inspector analyzes DCE/RPC over SMB.

  - **DCE TCP inspector**: The DCE TCP inspector analyzes the DCE/RPC protocol over TCP.

- **FTP client inspector: File Transfer Protocol (FTP)** is a protocol for transferring files from a server to a client. FTP works over TCP and usually operates on TCP ports 20 and 21. The FTP client inspector analyzes the FTP client commands and works in conjunction with the FTP server inspector.

- **FTP server inspector**: The FTP server inspector analyzes the FTP server response traffic and works with the FTP client inspector.

- **GTP Inspect inspector**: The GTP inspector analyzes the **GPRS Tunneling Protocol (GTP)** traffic.

- **Telnet inspector**: The Telnet inspector decodes and analyzes the Telnet protocol (usually over TCP port 23).

- **SIP inspector**: The **Session Initiation Protocol (SIP)** is a signaling protocol that enables **Voice over Internet Protocol (VoIP)**. The SIP inspector decodes and analyzes SIP.

- **SSH inspector**: The SSH inspector analyzes the **Secure Shell (SSH)** protocol.

Among the service inspectors, we see the special inspector, *wizard*, which is different from the others. As we saw, various service inspectors analyze specific protocols. In the typical case, the various application protocols use standard ports, but that is not necessary. Consider a TCP session where a certain client and server are talking HTTP on port 10000. In the case of such sessions, when a certain protocol is being used with a non-standard port, there is a need to identify the protocol and assign the traffic to a particular service inspector for analysis. The wizard inspector does a quick analysis to identify and hand over a particular network session to the most likely service inspector.

Next, let us look at the stream inspectors.

Stream inspectors

Stream inspectors primarily deal with session tracking for IP, TCP, UDP, and ICMP. The stream inspectors include the following:

- **stream**: There is also a *base* module that takes care of the configuration-related processing, as well as serving as the main entry point for stream-related processing. From there, the control is handed over to the appropriate stream handler based on the protocol.

- **stream_tcp**: TCP is a stateful, connection-oriented protocol. A TCP connection has two endpoints, usually referred to as a client and a server. For a meaningful analysis of TCP traffic by an IDS or IPS, the system has to maintain the state of both the client and server sides. This is done by the `stream_tcp` inspector.

  In addition, when the client and server send data back and forth, the data may be split across packet boundaries. Subsequently, there is a need to reassemble the transmitted data to a continuous data stream. This TCP reassembly is also done by the `stream_tcp` module.

- **stream_icmp**: When the protocol is ICMP, the stream base inspector forwards the packet to the `stream_icmp` inspector. Currently, this module only deals with ICMP unreachable messages. It ignores the remaining types.

- **stream_udp**: When the protocol is UDP, the stream base inspector forwards the packet to the `stream_udp` inspector. Since UDP is a connectionless protocol, the analysis required is much less than in the case of connection-oriented protocols.

- **stream_ip**: This inspector tracks IP (network) flows and enables session and flow tracking. The `stream_ip` inspector also deals with IP defragmentation (or IP fragmentation reassembly). IP fragmentation is a mechanism where an IP datagram (or packet) is split into multiple datagrams (or packets); this is done when the IP datagram size exceeds the data link layer's **Maximum Transmission Unit (MTU)**.

In the previous sections, we discussed three types of inspectors, namely, network, service, and stream inspectors, which were grouped based on the type of traffic they analyze. Besides these, there are two special inspectors: wizard and binder. These have a special role among the inspectors, namely, to direct the network traffic to the right inspector for analysis. We will delve more into these inspectors in the next section.

Wizard and binder inspectors

The wizard and binder inspectors are two special inspectors used by Snort 3. They determine the most appropriate service inspector to process a given flow and direct all the traffic pertaining to that flow to the determined inspector. These two inspectors have their special roles.

Wizard

The wizard inspector implements a port-independent protocol identification. Let's say that an endpoint machine runs an HTTP server on port 1000, and a client connects to that server on port 1000 using a browser and has an HTTP conversation. When analyzing this network traffic, the wizard inspector is designed to identify that the session uses HTTP, irrespective of the fact that port 1000 is not listed as one of the ports in the HTTP inspector configuration. Subsequently, the flow will be directed to the HTTP inspector for analysis.

The wizard's goal is to make its determination as quickly as possible so that the flow is analyzed by a service inspector. The determination by the wizard inspector can be inaccurate. A more accurate determination will be made by the relevant service inspector and the **Application Identification (AppID)** module. For identification, the wizard uses a set of patterns called hexes, spells, and curses.

Binder

The binder inspector applies a combination of conditions on the session to direct it to an inspector module for analysis. This is best illustrated by an example:

```
{ when = { proto = 'udp', ports = '53', role='server' },  use = { type
= 'dns' } }
```

The binder inspector configuration will include such rules. The when clause has three conditions, namely, the protocol is udp, the server port is 53, and the role is server. The server port is the destination port in a UDP packet. The use clause specifies which inspector to use. In this case, it is the DNS inspector.

In short, the wizard and binder inspectors are special and different from other inspector modules. They help in binding and/or connecting the traffic to the most appropriate service inspectors.

Summary

This chapter discussed inspectors, a crucial component of Snort 3. The inspector modules perform in-depth analyses of various network protocols and enable extensive rule-matching processes.

We discussed the role of inspectors and looked at the types of inspectors, and the various inspectors that fall under each type. We also covered two special inspectors – Binder and Wizard – and discussed their functionalities.

In the next chapter, we will examine stream inspectors in detail.

9
Stream Inspectors

Network traffic is comprised of packets or frames, which are the fundamental units of data transmission. These packets originate from a source endpoint and are transmitted to one or more destination endpoints. However, the individual packets are usually part of something bigger. For instance, when a web server communicates with a browser using HTTP, the data is divided into manageable sizes and sent as packets across the network. To conduct meaningful analysis, it is necessary to examine the data units of the underlying protocol, such as HTTP, rather than focusing solely on individual packets.

The analysis of network traffic becomes more complex due to the presence of numerous servers and clients concurrently exchanging data. When received by an IDS or IPS, the individual packets from different sources can become interspersed. Therefore, it becomes crucial for the IDS/IPS to accurately group these packets based on the relevant TCP connections and reconstruct the appropriate entities, such as HTTP data units, to facilitate further analysis and detection. This type of system is commonly referred to as a stateful network IDS or IPS.

In Snort 3, the stream inspector tracks and maintains the state of ongoing network connections. The module organizes and interprets the received packets, ensuring that the analysis and detection processes can be effectively performed. By understanding the relationships between packets within a connection, the system can reconstruct higher-level data units, such as complete HTTP requests or responses, for comprehensive inspection and threat identification.

In this chapter, we will discuss the stateful capability of Snort 3. We will discuss important terms such as flows, sessions, and streams, which are relevant to how Snort performs the stateful analysis. At a high level, we will discuss the following in this chapter:

- Relevant protocols for the stream inspector
- The stream inspectors

Relevant protocols for the stream inspector

The relevant protocols that are processed by a *stream* inspector are **Internet Protocol (IP)**, **User Datagram Protocol (UDP)**, **Internet Control Message Protocol (ICMP)**, and **Transmission Control Protocol (TCP)**. In this section, we will discuss some basics regarding these protocols. If you are well-versed in the details of the IP, ICMP, TCP, and UDP protocols, feel free to skip this section.

IP

Of these protocols, IP is a layer 3 (network layer) protocol according to the OSI model. This protocol deals with the delivery of network datagrams from a source to a destination, as specified in the IP header. The protocol follows a *best-effort* strategy, in which there is no guarantee that the datagram will reach the destination, or that the datagram will reach the destination error-free. If a source sends several datagrams to a particular destination using IP, there is no guarantee that all the datagrams will go via the same route (unless certain IP options are used such as **Loose Source Record Route (LSRR)** and **Strict Source Record Route (SSRR)**).

The IP portion of the packet starts with the IP header. The IP header consists of a minimum of 20 bytes up to a maximum of 60 bytes. The exact length of the IP header is encoded within the header itself. Let's take a look at the IP header next:

```
 0                   1                   2                   3
 0 1 2 3 4 5 6 7 8 9 0 1 2 3 4 5 6 7 8 9 0 1 2 3 4 5 6 7 8 9 0 1
+-+-+-+-+-+-+-+-+-+-+-+-+-+-+-+-+-+-+-+-+-+-+-+-+-+-+-+-+-+-+-+-+
|Version|  IHL  |Type of Service|          Total Length         |
+-+-+-+-+-+-+-+-+-+-+-+-+-+-+-+-+-+-+-+-+-+-+-+-+-+-+-+-+-+-+-+-+
|         Identification        |Flags|      Fragment Offset    |
+-+-+-+-+-+-+-+-+-+-+-+-+-+-+-+-+-+-+-+-+-+-+-+-+-+-+-+-+-+-+-+-+
|  Time to Live |    Protocol   |         Header Checksum        |
+-+-+-+-+-+-+-+-+-+-+-+-+-+-+-+-+-+-+-+-+-+-+-+-+-+-+-+-+-+-+-+-+
|                       Source Address                          |
+-+-+-+-+-+-+-+-+-+-+-+-+-+-+-+-+-+-+-+-+-+-+-+-+-+-+-+-+-+-+-+-+
|                    Destination Address                        |
+-+-+-+-+-+-+-+-+-+-+-+-+-+-+-+-+-+-+-+-+-+-+-+-+-+-+-+-+-+-+-+-+
|                       Options                 |    Padding    |
+-+-+-+-+-+-+-+-+-+-+-+-+-+-+-+-+-+-+-+-+-+-+-+-+-+-+-+-+-+-+-+-+
```

Figure 9.1 – IP header

Figure 9.1 shows the IP header. The length of the IP header is specified using 4 bits and the field is called IHL (which stands for **IP Header Length**). The value specified in this field has to be multiplied by 4 in order to get the header length. For example, a value of 5 for IHL would indicate that the IP header is 20 bytes long.

It can be noted that the `Total Length` field is 16 bits. Therefore, the maximum size of an IP datagram is 65,535 bytes.

> **Best-effort delivery**
>
> Best effort is a term often used in network protocol discussion, and it means that the protocol or program attempts the delivery of the message, while not attempting to track successful delivery or retransmissions in case of failures. When the underlying protocol uses the best-effort strategy, the application program or the upper layer using the service has to implement any necessary steps for tracking the success of the delivery or subsequent retransmissions in cases of failures.

IP fragmentation and reassembly

The layer 2 protocol typically has an upper limit for the data frame. For example, for Ethernet, the maximum frame size is 1514 bytes. This upper limit of layer 2 is called the maximum frame size. Correspondingly, there is a maximum size for an IP datagram that will fit the maximum-sized layer 2 frame. This is called the **Maximum Transmission Unit** (**MTU**). For Ethernet, the MTU is 1,500 bytes.

If layer 4 requests IP to transmit data greater than the MTU, the IP layer will have to divide the data into MTU-sized chunks and then transmit them. For example, if the data is 4,000 bytes, IP will divide the data into two 1,500-byte chunks and a 1,000-byte chunk. These chunks are called IP fragments. The resultant IP fragments will be as follows:

- **Fragment 1**: Fragment size 1,500, **More fragment** (**MF**) flag set, Fragment offset 0 (IP identifier 19746).

- **Fragment 2**: Fragment size 1,500, MF flag set, Fragment offset 1,500 (IP identifier 19746).

- **Fragment 3**: Fragment size 1,000, MF flag not set, Fragment offset 3,000 (IP identifier 19746).

The receiving end's IP layer will collect fragments based on the IP identifier, and when it has received all the fragments, it will reassemble them to create the intended IP datagram with an identifier value equal to 19746 and having 4,000 bytes.

ICMP

ICMP is a special protocol when it comes to OSI layering. Since it works closely with IP, it is considered a layer 3 protocol. However, it is encapsulated within an IP header, similar to a layer 4 protocol (such as UDP and TCP). ICMP deals with control messages in the operation of IP, and UDP. ICMP, like IP and UDP, is a connectionless protocol and differs from TCP, which is a connection-oriented protocol.

The following table lists some basic characteristics of IP, UDP, and ICMP as compared to TCP:

IP, UDP, and ICMP	TCP
Connectionless, stateless	Connection-oriented, stateful
Datagram-based	Stream-based
Unreliable, best-effort	Reliable

Table 9.1 – Differences between IP, UDP, ICMP, and TCP

UDP and TCP are layer 4 protocols according to the OSI model. They are used by applications or functionality that exist at layer 5 or above.

TCP

TCP is a connection-oriented protocol designed for dependable and sequential data transfer between two endpoints, typically a client and a server. It incorporates features such as reliable delivery, data ordering, and error checking. TCP also includes a flow control mechanism to regulate the rate of data transmission for efficient performance. The protocol utilizes specific messages to establish and terminate virtual connections, as well as to acknowledge the receipt of transmitted data. However, these advantageous features incur additional overhead due to the inclusion of extra messages.

In TCP, the process of connection establishment and termination is crucial for establishing a reliable communication channel between two endpoints. Here's an explanation of connection establishment and teardown in TCP.

Connection establishment

The client and server endpoints of TCP exchange certain control messages so that the connection is *established*. The first three messages (typical case) of the TCP connection that are sent for connection establishment are the SYN packet (from client to server), SYN-ACK (from server to client), and ACK (client to server); together, these constitute the three-way handshake of TCP:

1. SYN: The process begins with the initiating party (client) sending a SYN (**synchronize**) segment to the receiving party (server). The SYN segment carries an initial sequence number for establishing synchronization.

2. SYN-ACK: Upon receiving the SYN segment, the server responds with a SYN-ACK (**synchronize-acknowledge**) segment. This segment acknowledges the client's SYN segment, includes its own initial sequence number, and confirms the synchronization.

3. ACK: Finally, the client sends an ACK (**acknowledge**) segment to acknowledge the server's SYN-ACK segment. This completes the three-way handshake, and the connection is established. From this point on, both parties can transmit data using TCP.

In the typical case, the client starts from the CLOSED state and transitions through the SYN_SENT state to reach the ESTABLISHED state. Similarly, the server starts at the LISTEN state and then transitions through the SYN_RCVD state to reach the ESTABLISHED state.

Data transfer phase

The data transfer between the client and server happens only after the TCP connection is established. The data is transferred as a stream of data; each byte transferred is associated with a sequence number. There are two data streams – client to server, and server to client. The three-way handshake process establishes a set of sequence numbers (known as the initial sequence number) for both the client-to-server data stream and the server-to-client data stream. The data is made available to the endpoint in the same order (by sequence number) that the client sent it. This TCP layer at the receiving endpoint handles out-of-ordering, retransmissions, and overlaps, and provides the data as a stream; this functionality is called TCP stream reassembly. Once either side receives data from the other side, it sends an acknowledgment packet that confirms receipt. This ACK packet will acknowledge the sequence number of the last received byte of data.

Let's discuss the data transfer between client and server, and how the TCP ACK is used to acknowledge the receipt of data using a figure.

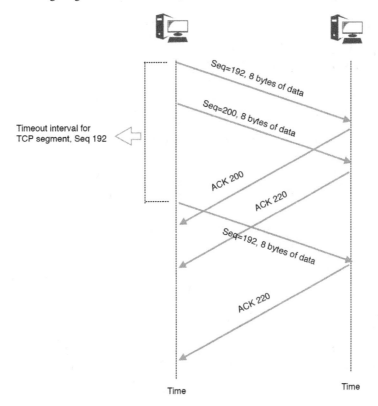

Figure 9.2 – TCP data transfer phase

In *Figure 9.2*, the initial sequence number for the client-to-server data stream is 192. The first data transfer is for 8 bytes. The sequence number for the last byte would be 199. Therefore, the server sends the ACK message that specifies 200, which means the server has received sequence number 199 and can receive 200 onward.

When the endpoint does not receive the acknowledgment message for the transferred data within a certain timeout period, it will resend the data (as seen in *Figure 9.2*). This is part of TCP so as to ensure reliable delivery.

Connection teardown

When the communication session is complete, either party can initiate the termination process:

1. FIN: The initiating party sends a FIN (**finish**) segment to the other party, indicating its intention to terminate the connection.

2. ACK: The receiving party acknowledges the FIN segment with an ACK segment.

3. FIN: In response, the receiving party may also initiate termination by sending a FIN segment.

4. ACK: The initiating party acknowledges the FIN segment with an ACK segment.

This four-way handshake ensures that both parties agree to terminate the connection. After the exchange of FIN and ACK segments, the connection is closed. However, it's important to note that TCP follows a reliable delivery mechanism, so even after the termination, any remaining data might still be delivered and acknowledged before the connection fully closes.

The connection establishment and teardown processes in TCP ensure a controlled and reliable communication channel between the client and server, providing ordered and error-checked data transfer.

The connection establishment, data transfer phase, and connection teardown phase are described using a TCP state diagram (shown in *Figure 9.3*).

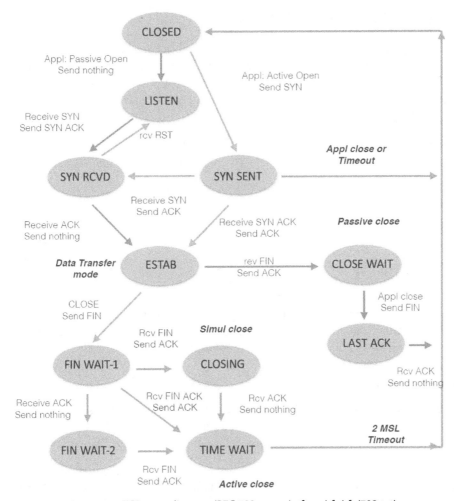

Figure 9.3 – TCP state diagram (RFC 793, www.ietf.org/rfc/rfc/793.txt)

Figure 9.3 shows the state diagram for TCP. The state transitions for the client are shown by the green arrows, and the state transitions for the server are shown by the blue arrows.

Flow control

TCP implements basic flow control to control or limit the rate at which the sender (client or server) sends data to the receiver side. This is implemented using a sliding window approach. The receiver advertises a TCP window to the sender indicating how many bytes of data can be sent by the sender without receiving an ACK. By reducing the TCP window to a small value, the receiver can slow down the sender, and by increasing the window to a large value, the receiver can tell the client that it can increase the data transfer rate. If the receiver receives data outside the advertised window, that data is rejected.

Out-of-order and overlapping TCP segments

TCP works on top of the network layer, which is IP in most cases. As mentioned earlier, IP is an unreliable (best-effort) protocol. This means that IP datagrams can get lost or can be corrupted. In addition, the consecutive IP datagrams need not traverse the same end-to-end path from source to destination. Due to these factors, the TCP data that is sent using IP may arrive at the destination in a different order of sequence numbers.

However, since each byte of the TCP data segment is associated with a sequence number, the server can arrange the segments in the correct order before passing them to the upper layer or application.

Due to the retransmission of data and the sliding window mechanism, there are scenarios that may result in overlapping TCP segments.

In the next section, we will look at another transport layer protocol that is like TCP but is different in many ways.

UDP

UDP is a layer 4 (transport layer) protocol that operates over the IP layer. Unlike TCP, UDP is a connectionless, stateless, and unreliable protocol. This means that UDP does not involve establishing or maintaining any connection information or state information. The protocol also does not include any mechanism to address packet loss or packet corruption. It has significantly less overhead compared to TCP and lacks flow control.

These characteristics make UDP ideal for quick and low-latency message transfer, where reliability is not critical. The ideal example usage scenario is in the field of real-time multimedia streaming and online gaming. UDP is the transport layer protocol for protocols such as **Domain Name System (DNS)** and **Dynamic Host Configuration Protocol (DHCP)**.

In the next section, let us discuss briefly the concept of flow as this is relevant for stream inspectors.

Flow

RFC 3917 defines *flow* as follows:

> *"A flow is defined as a set of IP packets passing an observation point in the network during a certain time interval. All packets belonging to a particular flow have a set of common properties."*

The common properties that pertain to a flow are called a flow key in Snort 3. The flow key differs between TCP, ICMP, and IP (UDP has the same flow key as TCP).

In the case of TCP, the flow key consists of the source IP address, source port, destination IP address, destination port, protocol value, VLAN ID, and MPLS ID (when VLAN and MPLS are not applicable, the VLAN ID and MPLS ID will be set to 0).

In the case of ICMP, the flow key parameters will include source and destination IP address (source and destination ports are not meaningful for ICMP), ICMP type, protocol, VLAN ID, and MPLS ID.

For IP, the flow key includes source and destination IP addresses, IP identifiers, VLAN ID, and MPLS ID.

Upon analyzing a new flow, Snort assigns a flow object data structure to track the flow. The flow object is stored in a cache that uses the flow key for look-up. Snort also attempts to identify the protocol associated with the flow and assign it to the appropriate service inspector.

In the next section, let us discuss the different stream inspectors and the details pertaining to each type.

The stream inspectors

As we saw, the protocols – IP, UDP, ICMP, and TCP – have complicated scenarios that need to be handled precisely to ensure accurate analysis, avoid false positives, and evasion attempts. The inspectors maintain the necessary connection management, accurate state tracking, fragmentation reassembly, and stream reassembly so that the rest of the analysis (layer 5 and above) can be performed accurately.

Let us look closer at each inspector and see the important functions that they perform. We will start with the stream_ip inspector.

stream_ip

The stream_ip inspector does session tracking for IP flows. In addition, this inspector analyzes fragmented IP datagrams and does the IP fragmentation reassembly operation.

Although IP is a stateless protocol, from a session tracking point of view, there is a set of states that are assigned to the IP flow. Upon seeing traffic for a particular IP flow in one direction, the SSNFLAG_SEEN_CLIENT flag is set for that flow. Similarly, upon seeing traffic for the flow in the opposite direction, the SSNFLAG_SEEN_SERVER flag is set. When both SSNFLAG_SEEN_CLIENT and SSNFLAG_SEEN_SERVER are set, the flow is considered SSNFLAG_ESTABLISHED.

Configuration

The configuration options for the stream_ip inspector include the following options (the configuration options for each module can be listed using the help command available with Snort):

```
snort3 --help-module stream_ip
```

Let us briefly discuss the various configuration options available for `snort_ip`:

- `max_frags`: Upon analyzing an IP fragment, Snort creates a flow and tracks all the fragments with the same flow key. However, using the `max_frags` parameter, Snort limits the number of simultaneous fragmentations that are tracked. This is to avoid resource attacks (DoS attacks) against the Snort 3 IDS/IPS:

```
int stream_ip.max_frags = 8192: maximum number of simultaneous
fragments being tracked { 1:max32 }
```

- `max_overlaps`: IP fragments may overlap, and the fragmentation reassembly process will correctly reassemble the datagram from all the fragmented chunks, even in the presence of overlap. However, too many overlap cases may indicate evasion attempts or malicious behavior in general. The `max_overlaps` parameter limits the number of overlaps per flow:

```
int stream_ip.max_overlaps = 0: maximum allowed overlaps per
datagram; 0 is unlimited { 0:max32 }
```

- `min_frag_length`: Fragmented packets that are too small may also indicate evasion attacks. The configuration parameter called `min_frag_length` sets a minimum length for fragmented IP packets to be compared against to detect possible evasion attempts:

```
int stream_ip.min_frag_length = 0: alert if fragment length is
below this limit before or after trimming { 0:65535 }
```

- `min_ttl`: Fragmented packets that have very low **Time-To-Live** (**TTL**) values may also indicate evasion attacks. The `min_ttl` parameter sets a minimum TTL value for fragmented IP packets to be compared against to detect possible evasion attempts:

```
int stream_ip.min_ttl = 1: discard fragments with TTL below the
minimum { 1:255 }
```

- `policy`: When there is IP fragmentation overlap, various OSs use different approaches regarding which data to use in the reassembled IP datagram. The `policy` parameter specifies which approach to follow. `first` means that in case of overlap, use the data that arrived first. Parameter values such as `linux` or `bsd` specify which approach to follow when it comes to IP fragmentation reassembly:

```
enum stream_ip.policy = 'linux': fragment reassembly policy {
'first' | 'linux' | 'bsd' | 'bsd_right' | 'last' | 'windows' |
'solaris' }
```

- `session_timeout`: IP is stateless and does not have a teardown phase or establishing phase. If there is no activity in the flow, it will be timed out after a timeout period. This timeout period is configurable using the `session_timeout` variable:

```
stream_ip.session_timeout = 60: session tracking timeout {
1:max31 }
```

In the next section, let us take a look at the various alerts that are generated by the stream_ip inspector.

Alerts from the stream_ip inspector

The stream_ip inspector can detect and alert on a set of conditions, as shown in the following table:

gid:sid	Alert message
123:1	(stream_ip) inconsistent IP options on fragmented packets
123:2	(stream_ip) teardrop attack
123:3	(stream_ip) short fragment, possible DoS attempt
123:4	(stream_ip) fragment packet ends after defragmented packet
123:5	(stream_ip) zero-byte fragment packet
123:6	(stream_ip) bad fragment size, packet size is negative
123:7	(stream_ip) bad fragment size, packet size is greater than 65536
123:8	(stream_ip) fragmentation overlap
123:11	(stream_ip) TTL value less than configured minimum, not used for reassembly
123:12	(stream_ip) excessive fragment overlap
123:13	(stream_ip) tiny fragment

Table 9.2 – Alerts generated by the stream_ip inspector

Every alert in Snort has a **Generator ID** (gid) and a **Signature ID** (sid). gid is used to identify which module or component of Snort generated the alert, and sid specifies the specific condition. The gid and sid values for each of the alerts generated from the stream_ip module are listed in the first column, and the second column specifies the alert message.

Let us briefly discuss each of these alerts and the condition in which they trigger:

- **Inconsistent IP options on fragmented packets**: When an IP datagram is fragmented, all of the IP options of the original IP datagram should be copied to the first IP fragment, while only the relevant IP options of the original datagram need to be copied to all the subsequent fragmented IP datagrams. When any inconsistency in the IP options of the analyzed IP fragments is observed, Snort generates this alert.

- **Teardrop attack**: The teardrop attack involves the attacker sending malformed IP fragments to a victim's machine. Every IP fragment contains an offset, and also the fragment size. The IP fragmentation reassembly uses the offset and size to recreate the original IP datagram. When the offset and sizes of the various IP fragments are such that it results in fragmentation overlaps, and the victim's machine does not handle it correctly, it results in a crash and hence a **Denial of Service (DoS)** condition. Snort detects this attack and creates this alert.

- **Short fragment, possible DoS attempt**: All non-last fragments are supposed to end on 8-byte boundaries. When this does not happen, this alert is generated.

- **Fragment packet ends after defragmented packet**: This is an anomaly case where the end of the IP fragment being processed is greater than the calculated size of the defragmented IP packet.

- **Zero-byte fragment packet**: This alert is generated when an IP fragment is received with zero size. This is an abnormal case and should not be seen in normal traffic.

- **Bad fragment size, packet size is negative**: This alert is generated for an anomalous case where the end of the fragmentation is before the fragment offset, which can only happen if the packet size is negative – hence, abnormal.

- **Bad fragment size, packet size greater than 65536**: The defragmentation cannot create an IP packet that is greater than 65,535, which is the size of the largest IP datagram. This alert is generated when such a case is encountered.

- **Fragmentation overlap**: IP fragmentation overlap may occur in rare conditions, but Snort can generate an alert for this condition.

- **TTL value less than the configured minimum, not used for reassembly**: This alert is generated when the TTL value of a fragment is less than the configured minimum value. This alert is useful to detect evasion efforts.

- **Excessive fragment overlap**: IP fragmentation overlap may occur in rare conditions, but an excessive number of overlapping fragments may indicate an intentional effort (attack).

- **Tiny fragment**: This alert is generated when the received IP fragment size is less than the configured `min_fragment_length`.

Flow tracking and IP fragmentation reassembly are critical functionality for Snort's operation. These are performed by the `stream_ip` inspector. Next, let us look at a layer 4 inspector – `stream_udp`.

stream_udp

The `stream_udp` inspector is a lightweight module. The main task that it does is to track UDP flows. Like the `stream_ip` inspector, although UDP is a stateless protocol, from a session tracking point of view, there is a set of states that are assigned to the UDP flow. Upon seeing traffic for a particular IP flow in one direction, the `SSNFLAG_SEEN_SENDER` flag is set for that flow. Similarly, upon seeing traffic for the flow in the opposite direction, the `SSNFLAG_SEEN_RESPONDER` flag is set. When both `SSNFLAG_SEEN_SENDER` and `SSNFLAG_SEEN_RESPONDER` are set, the flow is considered `SSNFLAG_ESTABLISHED`.

Configuration

There is only a single configuration parameter for the `stream_udp` inspector:

- `session_timeout`: UDP is stateless and does not have a teardown phase or establishing phase. When there is no activity in the flow, it will be timed out after a timeout period. This timeout period is configurable using the `session_timeout` variable:

    ```
    int stream_udp.session_timeout = 30: session tracking timeout {
    1:max31 }
    ```

There are no alerts that are generated from within the `stream_udp` inspector.

UDP is a lightweight layer 4 protocol. Correspondingly, the `stream_udp` inspector is also pretty lightweight. Next, we will look into the `stream_icmp` inspector that analyzes ICMP.

stream_icmp

The `stream_icmp` inspector is also a lightweight module. It tracks ICMP flows. ICMP is a stateless protocol; however, from a session tracking point of view, there is a set of states that are assigned to the ICMP flow. Upon seeing traffic for a particular IP flow in one direction, the `SSNFLAG_SEEN_SENDER` flag is set for that flow. Similarly, upon seeing traffic for the flow in the opposite direction, the `SSNFLAG_ESTABLISHED` is set.

Among all the ICMP types, the only message type that the `stream_icmp` inspector handles is *ICMP Destination Unreachable*.

Configuration

There is only a single configuration parameter for the `stream_icmp` inspector:

- `session_timeout`: ICMP is stateless and does not have a teardown phase or establishing phase. When there is no activity in the flow, it will be timed out after a timeout period. This timeout period is configurable using the `session_timeout` variable:

```
int stream_icmp.session_timeout = 30: session tracking timeout {
1:max31 }
```

There are no alerts that are generated from within the `stream_icmp` inspector.

Similar to UDP, ICMP is a lightweight protocol. In this section, we discussed the ICMP inspector and discussed its configuration. In the next section, we will discuss the `stream_tcp` inspector.

stream_tcp

TCP is a connection-oriented and stateful protocol. The `stream_tcp` inspector performs a stateful analysis of TCP, tracks the flow, and does TCP stream reassembly so that the rest of the analysis is not just packet-based, but stream-based (which enables protocol data unit-based analysis). To perform the TCP stream reassembly, `stream_tcp` also handles out-of-order TCP segments, TCP retransmissions, and TCP overlap scenarios.

Let's look at an example to understand the TCP stream reassembly operation. Consider a typical HTTP session in which the client sends the following HTTP request:

```
GET /index.html HTTP/1.1
Host: test.edu
Connection: close
```

We also have a test signature that creates a Snort alert when the preceding HTTP request is processed:

```
alert tcp any any -> any any (msg:"Request to /index.html detected
- Test"; flow:established; http_method; content:"GET"; http_uri;
content:"|2F|index"; sid:123; priority:3;)
```

Now, consider the case where the data is not contained in a single TCP segment, but has been segmented into multiple TCP segments. The network bytes when the preceding HTTP request is sent over the wire as 2-byte TCP segments are as follows:

```
03:18:23.047163 IP 192.168.0.1.39484 > 192.168.0.2.80: Flags [P.],
seq 0:2, ack 1, win 260, options [nop,nop,TS val 1092427803 ecr
260077651], length 2: HTTP
        0x0000:  4500 0036 26b9 4000 4006 92b5 c0a8
0001   E..6&.@.@.......
        0x0010:  c0a8 0002 9a3c 0050 9cc5 1b64 13e6
3b15   .....<.P...d..;.
```

```
          0x0020:  8018 0104 2259 0000 0101 080a 411d
201b  ...."Y......A...
          0x0030:  0f80 7853 4745                       ..xSGE
03:18:23.047176 IP 192.168.0.1.39484 > 192.168.0.2.80: Flags [P.],
seq 2:4, ack 1, win 260, options [nop,nop,TS val 1092427803 ecr
260077651], length 2: HTTP
          0x0000:  4500 0036 26ba 4000 4006 92b4 c0a8
0001  E..6&.@.@.......
          0x0010:  c0a8 0002 9a3c 0050 9cc5 1b66 13e6
3b15  .....<.P...f..;.
          0x0020:  8018 0104 157c 0000 0101 080a 411d
201b  .....|......A...
          0x0030:  0f80 7853 5420                       ..xST.
03:18:23.047184 IP 192.168.0.1.39484 > 192.168.0.2.80: Flags [P.],
seq 4:6, ack 1, win 260, options [nop,nop,TS val 1092427803 ecr
260077651], length 2: HTTP
          0x0000:  4500 0036 26bb 4000 4006 92b3 c0a8
0001  E..6&.@.@.......
          0x0010:  c0a8 0002 9a3c 0050 9cc5 1b68 13e6
3b15  .....<.P...h..;.
          0x0020:  8018 0104 3a31 0000 0101 080a 411d
201b  ....:1......A...
          0x0030:  0f80 7853 2f69                       ..xS/i
03:18:23.047190 IP 192.168.0.1.39484 > 192.168.0.2.80: Flags [P.],
seq 6:8, ack 1, win 260, options [nop,nop,TS val 1092427803 ecr
260077651], length 2: HTTP
          0x0000:  4500 0036 26bc 4000 4006 92b2 c0a8
0001  E..6&.@.@.......
          0x0010:  c0a8 0002 9a3c 0050 9cc5 1b6a 13e6
3b15  .....<.P...j..;.
          0x0020:  8018 0104 fb33 0000 0101 080a 411d
201b  .....3......A...
          0x0030:  0f80 7853 6e64                       ..xSnd
03:18:23.047196 IP 192.168.0.1.39484 > 192.168.0.2.80: Flags [P.],
seq 8:10, ack 1, win 260, options [nop,nop,TS val 1092427803 ecr
260077651], length 2: HTTP
          0x0000:  4500 0036 26bd 4000 4006 92b1 c0a8
0001  E..6&.@.@.......
          0x0010:  c0a8 0002 9a3c 0050 9cc5 1b6c 13e6
3b15  .....<.P...l..;.
          0x0020:  8018 0104 041e 0000 0101 080a 411d
201b  ............A...
          0x0030:  0f80 7853 6578                       ..xSex
03:18:23.047200 IP 192.168.0.1.39484 > 192.168.0.2.80: Flags [P.],
seq 10:12, ack 1, win 260, options [nop,nop,TS val 1092427803 ecr
260077651], length 2: HTTP
          0x0000:  4500 0036 26be 4000 4006 92b0 c0a8
0001  E..6&.@.@.......
          0x0010:  c0a8 0002 9a3c 0050 9cc5 1b6e 13e6
```

```
3b15    .....<.P...n..;.
        0x0020:  8018 0104 3b2c 0000 0101 080a 411d
201b    ....;,......A...
        0x0030:  0f80 7853 2e68                          ..xS.h
03:18:23.047217 IP 192.168.0.1.39484 > 192.168.0.2.80: Flags [P.],
seq 12:14, ack 1, win 260, options [nop,nop,TS val 1092427803 ecr
260077651], length 2: HTTP
        0x0000:  4500 0036 26bf 4000 4006 92af c0a8
0001    E..6&.@.@.......
        0x0010:  c0a8 0002 9a3c 0050 9cc5 1b70 13e6
3b15    .....<.P...p..;.
        0x0020:  8018 0104 f524 0000 0101 080a 411d
201b    .....$......A...
        0x0030:  0f80 7853 746d                          ..xStm
03:18:23.047224 IP 192.168.0.1.39484 > 192.168.0.2.80: Flags [P.],
seq 14:16, ack 1, win 260, options [nop,nop,TS val 1092427803 ecr
260077651], length 2: HTTP
        0x0000:  4500 0036 26c0 4000 4006 92ae c0a8
0001    E..6&.@.@.......
        0x0010:  c0a8 0002 9a3c 0050 9cc5 1b72 13e6
3b15    .....<.P...r..;.
        0x0020:  8018 0104 fd6f 0000 0101 080a 411d
201b    .....o......A...
        0x0030:  0f80 7853 6c20                          ..xSl.
03:18:23.047230 IP 192.168.0.1.39484 > 192.168.0.2.80: Flags [P.],
seq 16:18, ack 1, win 260, options [nop,nop,TS val 1092427803 ecr
260077651], length 2: HTTP
        0x0000:  4500 0036 26c1 4000 4006 92ad c0a8
0001    E..6&.@.@.......
        0x0010:  c0a8 0002 9a3c 0050 9cc5 1b74 13e6
3b15    .....<.P...t..;.
        0x0020:  8018 0104 213a 0000 0101 080a 411d
201b    ....!:......A...
        0x0030:  0f80 7853 4854                          ..xSHT
03:18:23.047236 IP 192.168.0.1.39484 > 192.168.0.2.80: Flags [P.],
seq 18:20, ack 1, win 260, options [nop,nop,TS val 1092427803 ecr
260077651], length 2: HTTP
        0x0000:  4500 0036 26c2 4000 4006 92ac c0a8
0001    E..6&.@.@.......
        0x0010:  c0a8 0002 9a3c 0050 9cc5 1b76 13e6
3b15    .....<.P...v..;.
        0x0020:  8018 0104 153c 0000 0101 080a 411d
201b    .....<......A...
        0x0030:  0f80 7853 5450                          ..xSTP
03:18:23.047242 IP 192.168.0.1.39484 > 192.168.0.2.80: Flags [P.],
seq 20:22, ack 1, win 260, options [nop,nop,TS val 1092427803 ecr
260077651], length 2: HTTP
        0x0000:  4500 0036 26c3 4000 4006 92ab c0a8
0001    E..6&.@.@.......
```

```
          0x0010:  c0a8 0002 9a3c 0050 9cc5 1b78 13e6
3b15  .....<.P...x..;.
          0x0020:  8018 0104 3a59 0000 0101 080a 411d
201b  ....:Y.......A...
          0x0030:  0f80 7853 2f31                        ..xS/1
03:18:23.047248 IP 192.168.0.1.39484 > 192.168.0.2.80: Flags [P.],
seq 22:24, ack 1, win 260, options [nop,nop,TS val 1092427803 ecr
260077651], length 2: HTTP
          0x0000:  4500 0036 26c4 4000 4006 92aa c0a8
0001  E..6&.@.@.......
          0x0010:  c0a8 0002 9a3c 0050 9cc5 1b7a 13e6
3b15  .....<.P...z..;.
          0x0020:  8018 0104 3b57 0000 0101 080a 411d
201b  ....;W.......A...
          0x0030:  0f80 7853 2e31                        ..xS.1
03:18:23.047254 IP 192.168.0.1.39484 > 192.168.0.2.80: Flags [P.],
seq 24:26, ack 1, win 260, options [nop,nop,TS val 1092427803 ecr
260077651], length 2: HTTP
          0x0000:  4500 0036 26c5 4000 4006 92a9 c0a8
0001  E..6&.@.@.......
          0x0010:  c0a8 0002 9a3c 0050 9cc5 1b7c 13e6
3b15  .....<.P...|..;.
          0x0020:  8018 0104 5c7c 0000 0101 080a 411d
201b  ....\|.......A...
          0x0030:  0f80 7853 0d0a                        ..xS..
```

The stream reassembly functionality of stream_tcp performs the TCP reassembly operation.

For this test, we set the show_rebuilt_packets option in the stream_tcp module configuration, as given in the following code. This option causes Snort to dump out the rebuilt TCP packets to the stdout:

```
stream_tcp =
{
    show_rebuilt_packets = true
}
```

The Snort stdout printed out the following:

```
07/02-03:18:23.047163 TCP 192.168.0.1:39484 -> 192.168.0.2:80
http_inspect.stream_tcp[24]:
47 45 54 20 2F 69 6E 64   65 78 2E 68 74 6D 6C 20  GET /ind ex.html
48 54 54 50 2F 31 2E 31                            HTTP/1.1


07/02-03:18:23.047260 TCP 192.168.0.1:39484 -> 192.168.0.2:80
http_inspect.stream_tcp[33]:
48 6F 73 74 3A 20 74 65   73 74 2E 65 64 75 0D 0A  Host: te st.edu..
43 6F 6E 6E 65 63 74 69   6F 6E 3A 20 63 6C 6F 73  Connecti on: clos
65                                                 e
```

Here, we see how the `stream_tcp` module reassembled the 2-byte TCP segments and created two rebuilt packets. The test also triggers the following alert (based on the signature mentioned previously):

```
07/02-03:18:23.047260 [**] [1:123:0] "Request to /index.html detected
- Test" [**] [Priority: 3] {TCP} 192.168.0.1:39484 -> 192.168.0.2:80
```

Now, let us take a look at the various configuration settings for `stream_tcp`.

Configuration

The configuration settings for the `stream_tcp` module include the following options:

- `max_window`: TCP implements flow control using the advertised TCP window. The TCP window value in the TCP header is limited to `65535` since the TCP window field is 16 bits. The TCP window scaling option makes it possible to have a larger window. The maximum value supported is 1,073,725,440 (which is 65535 times $2^{**}14$). RFC1323 recommends that if a TCP window greater than this is advertised, then an error has to be logged and the maximum window size used as the value. `stream_tcp` also generates an alert in this scenario (the alert will be discussed in the next section):

  ```
  int stream_tcp.max_window = 0: maximum allowed TCP window {
  0:1073725440 }
  ```

- `overlap_limit`: TCP overlapping of data is not an unusual or abnormal case. However, too many overlapping segments may be unusual and could indicate the possibility of evasion attempts. Therefore, it is good to detect and/or limit the amount of overlapping data that can happen for a TCP session. Once this limit is reached, the inspector logs the following alert: **129:7 (stream_tcp) limit on number of overlapping TCP packets reached**. This configuration option sets the limit for the number of overlapping segments.

  ```
  int stream_tcp.overlap_limit = 0: maximum number of allowed
  overlapping segments per session { 0:max32 }
  ```

- `max_pdu`: The `stream_tcp` inspector does TCP stream reassembly; the TCP data bytes that are sent using separate IP datagrams are put together in the order of TCP sequence numbers (accounting for overlaps, retransmissions, and so on). In addition, this is done so as to create data blocks called **Protocol Data Units** (**PDUs**). This configuration option limits the size of the thus-created PDUs:

  ```
  int stream_tcp.max_pdu = 16384: maximum reassembled PDU size {
  1460:32768 }
  ```

- `no_ack`: The `no_ack` option is valid only in IPS mode. The TCP segments are processed and purged even without seeing an `ACK` from the receiving endpoint:

  ```
  bool stream_tcp.no_ack = false: received data is implicitly
  acked immediately
  ```

- `policy`: In the process of TCP reassembly, there are conditions such as TCP overlaps, TCP retransmissions, and so on, where various OSs use different approaches to pick which data to use in the reassembled TCP stream. The `policy` parameter specifies which approach to follow. `first` means that in case of overlap, use the data that arrived first. Parameter values such as `linux` or `bsd` specify which approach to follow when it comes to TCP reassembly:

```
enum stream_tcp.policy = 'bsd': determines operating system
characteristics like reassembly { 'first' | 'last' | 'linux' |
'old_linux' | 'bsd' | 'macos' | 'solaris' | 'irix' | 'hpux11' |
'hpux10' | 'windows' | 'win_2003' | 'vista' | 'proxy' }
```

- `reassemble_async`: This option, when enabled, tells Snort to queue TCP segment data before traffic is seen in both directions. This is useful in some special conditions where the traffic packets are not analyzed in the order they were sent, but rather, in an asynchronous fashion:

```
bool stream_tcp.reassemble_async = true: queue data for
reassembly before traffic is seen in both directions
```

- `require_3whs`: We saw the three-way handshake process for TCP connection establishment earlier in the chapter. This is the process whereby both endpoints (client and server) negotiate the **initial sequence number (ISN)** for both directions. The three-way handshake is an essential step to ensure reliable data transfer.

  Similarly, tracking the three-way handshake is an important step in analyzing the TCP connection and flow. However, as the IDS/IPS starts analysis, certain TCP connections may have already been in the ESTABLISHED state or may already be in the middle of the three-way handshake process. Such TCP connections are called midstream TCP sessions.

  The `require_3whs` parameter specifies a certain number of seconds, say, n seconds; Snort will be lenient in analyzing midstream TCP sessions for n seconds. After n seconds, when Snort encounters a new TCP session that seems to be in midstream, it will not analyze it. A value of -1 for `require_3whs` lets Snort analyze midstream TCP sessions always:

```
int stream_tcp.require_3whs = -1: don't track midstream sessions
after given seconds from start up; -1 tracks all { -1:max31 }
```

- `show_rebuilt_packets`: This is an option that is useful in troubleshooting. When this Boolean option is enabled, the `-X` option for Snort will display reassembled TCP pseudo packets also:

```
bool stream_tcp.show_rebuilt_packets = false: enable cmg like
output of reassembled packets
```

- `flush_factor`: This option enables a `flush` operation upon seeing a drop in segment size after a given number of non-decreasing segments:

```
int stream_tcp.flush_factor = 0: flush upon seeing a drop in
segment size after given number of non-decreasing segments {
0:65535 }
```

- `queue_limit.max_bytes`: The `stream_tcp` module maintains a queue for the TCP segments sent by the client to the server and another queue for the other direction. The TCP stream reassembly is done by combining the various TCP segments from these queues. Based on various conditions, these queues are purged periodically.

  This option – `queue_limit.max_bytes` – sets an upper limit for the amount of bytes stored in these queues. Once this limit is reached, subsequent data segments are not queued until the queue is purged and the byte count goes below the limit:

  ```
  int stream_tcp.queue_limit.max_bytes = 4194304: don't queue more
  than given bytes per session and direction, 0 = unlimited {
  0:max32 }
  ```

- `queue_limit.max_segments`: This option is similar to the previous option but differs in one aspect. Where the previous option puts a limit on the total bytes that are stored in the queue, this option puts a maximum limit on the number of TCP segments stored in the queue:

  ```
  int stream_tcp.queue_limit.max_segments = 3072: don't queue more
  than given segments per session and direction, 0 = unlimited {
  0:max32 }
  ```

- `small_segments.count` and `small_segments.maximum_size`: When an application uses TCP to send a large amount of data, the data is split into multiple TCP segments (based on the **Maximum Segment Size (MSS)**). In the case of Ethernet networks, the MSS is around 1,460 bytes. Certain applications such as telnet applications, which use TCP, send small-sized data. As a user types on the client side, the data of a few bytes is immediately sent to the other end. This is done for a better interactive feel for the user.

  So, depending on the applications being used and the environment, the presence of an unusual number of small TCP segments may indicate attempts to evade detection. For example, if TCP stream reassembly is not enabled, using small segments to split the attack traffic into several packets is an easy way to evade detection by the IDS.

  The two options – `small_segments.maximum_size` and `small_segments.count` – together can be used to detect such evasion attempts. The inspector will keep a count of the number of TCP segments that are below the specified `small_segments.maximum_size` value, and when the count is greater than the `small_segments.count` value, it generates an alert:

  ```
  int stream_tcp.small_segments.count = 0: number of consecutive
  (in the received order) TCP small segments considered to be
  excessive (129:12) { 0:2048 }
  int stream_tcp.small_segments.maximum_size = 0: minimum bytes
  for a TCP segment not to be considered small (129:12) { 0:2048 }
  ```

  Unless configured correctly with the knowledge of the environment, this can create a lot of false positives and noisy alerts.

- `session_timeout`: Usually, TCP connections end using the four-way handshake, as can be noted in *Figure 9.3*. TCP sessions may also be aborted by TCP reset segments. In some cases, TCP connections may be idle for a really long time, without being closed or aborted, and also without any data transfer. In such conditions, the session may be timed out by Snort. The `session_timeout` option specifies a time period in seconds after which an idle TCP connection will be timed out and cleared:

```
int stream_tcp.session_timeout = 180: session tracking timeout {
1:max31 }
```

- `track_only`: This option causes the `stream_tcp` inspector to perform only state tracking and disables TCP stream reassembly completely:

```
bool stream_tcp.track_only = false: disable reassembly if true
```

In the next section, let us take a look at the various alerts that are generated by the `stream_tcp` inspector.

Alerts from the stream_tcp inspector

The `stream_tcp` inspector identifies a set of abnormal or unusual scenarios that may indicate a malicious action or behavior. The various alerts generated from within the `stream_tcp` inspector are as follows:

gid:sid	Message
129:1	`(stream_tcp) SYN on established session`
129:2	`(stream_tcp) data on SYN packet`
129:3	`(stream_tcp) data sent on stream not accepting data`
129:4	`(stream_tcp) TCP timestamp is outside of the PAWS window`
129:5	`(stream_tcp) bad segment, adjusted size <= 0 (deprecated)`
129:6	`(stream_tcp) window size (after scaling) larger than policy allows`
129:7	`(stream_tcp) limit on number of overlapping TCP packets reached`
129:8	`(stream_tcp) data sent on stream after TCP reset sent`
129:9	`(stream_tcp) TCP client possibly hijacked, different ethernet address`

gid:sid	Message
129:10	(stream_tcp) TCP server possibly hijacked, different ethernet address
129:11	(stream_tcp) TCP data with no TCP flags set
129:12	(stream_tcp) consecutive TCP small segments exceeding threshold
129:13	(stream_tcp) 4-way handshake detected
129:14	(stream_tcp) TCP timestamp is missing
129:15	(stream_tcp) reset outside window
129:16	(stream_tcp) FIN number is greater than prior FIN
129:17	(stream_tcp) ACK number is greater than prior FIN
129:18	(stream_tcp) data sent on stream after TCP reset received
129:19	(stream_tcp) TCP window closed before receiving data
129:20	(stream_tcp) TCP session without 3-way handshake

Table 9.3 – Alerts generated by the stream_tcp inspector

stream_base

The stream_base inspector is the entry point for the stream evaluation of a packet. The stream_base inspector handles the stream global configuration management and the stream statistics management.

Configuration

The configuration settings for the stream_base module include the following options:

- ip_frags_only: This option tells the stream inspector to only process IP-fragmented packets:

    ```
    bool stream.ip_frags_only = false: don't process non-frag flows
    ```

- max_flows: Tracking a flow consumes memory. This configuration parameter sets a maximum limit for the simultaneous flows that are tracked by Snort:

    ```
    int stream.max_flows = 476288: maximum simultaneous flows
    tracked before pruning { 2:max32 }
    ```

- `pruning_timeout`: This configuration parameter sets a certain period in seconds; flows that are idle for this period of time are pruned:

```
int stream.pruning_timeout = 30: minimum inactive time before
being eligible for pruning { 1:max32 }
```

There are no alerts associated with the `stream_base` inspector.

Summary

In this chapter, we discussed the different stream inspectors. The stream inspectors are a set of modules that perform critical and fundamental functions in an IDS/IPS, such as flow tracking, IP defragmentation, and TCP reassembly. The stream inspectors included `stream_tcp`, `stream_ip`, `stream_udp`, and `stream_icmp`. We discussed the role and functionality of each of these inspector modules and also looked at the configuration parameters relevant to these inspectors. Finally, we discussed the alerts that are generated from them. The chapter presented the stream inspectors, their role, and their importance in the IDS/IPS functionality. This should enable you to tweak the stream inspector configuration as per the needs of your environment or setting.

In the next chapter, we will learn about the HTTP inspector in Snort 3.

10

HTTP Inspector

HTTP is one of the most prevalent protocols used over the internet. Recent statistics show that there are more than five billion internet users currently. There are more than 150 million .com domains, which is the most of any top-level domains. The total number of websites exceeds 2 billion; there were 300 million online shoppers in the US in 2023. Web traffic is at the core of the internet, and HTTP still plays a key role. Subsequently, a significant percentage of attacks and breaches occur via HTTP. According to the *2023 Data Breach Investigations Report* conducted by Verizon, 26% of breaches occurred via HTTP. In order to successfully detect malicious behavior and attacks over HTTP, the Snort system will have to decode the protocol and enable the identification of malicious and/or suspicious artifacts. The HTTP Inspector module performs the decoding and analysis of HTTP to enable such detection.

In this chapter, we will cover the following topics:

- Basics of HTTP
- HTTP inspector
- HTTP inspector configuration

Basics of HTTP

HTTP stands for **Hypertext Transfer Protocol**. HTTP is a plain text-based and stateless application protocol that operates on top of the TCP layer and is used for retrieving and delivering graphics, audio, video, plain text, and multimedia content.

The **Requests for Comments (RFC)** publication for HTTP/1.1 (RFC 9112) defines HTTP as follows:

> *"The Hypertext Transfer Protocol (HTTP) is a stateless application-level request/ response protocol that uses extensible semantics and self-descriptive messages for flexible interaction with network-based hypertext information systems."*

HTTP works using the client-server model and uses a request-response method. The web client is typically a browser, and it interacts with a web server using HTTP.

HTTP works over TCP as layer 4. Once the TCP handshake is complete, the client initiates by sending an HTTP request. In the next section, let us talk about the HTTP request.

HTTP request

As mentioned, HTTP follows a request-response approach. The HTTP client initiates the communication by sending the HTTP request, and the server responds with a response. The HTTP request that is sent from the HTTP client to the server has a particular syntax. The request contains a request line that specifies the resource that is accessed and the operation that is done to the resource. For example, a GET method is used to get the latest copy of a resource, such as a web page. In addition, the HTTP request will contain additional information that is sent as HTTP headers.

According to RFC 2616, the HTTP request has the following syntax:

```
Request-Line
 *(( general-header
 | request-header
 | entity-header ) CRLF)
  CRLF
  [ message-body ]
```

The HTTP request line contains the method, URI, and HTTP version, and has the following syntax (note that SP indicates a *space* character):

```
Method SP Request-URI SP HTTP-Version CRLF
```

The URI can have the following parts: scheme, host, TCP port, path, query, and fragment.

The protocol supports the following set of HTTP request methods: GET, POST, HEAD, PUT, DELETE, CONNECT, OPTIONS, TRACE, and PATCH. Each of the HTTP request methods has a specific purpose briefly described as follows:

- GET: The GET method, as the name suggests, retrieves specific data (as indicated by the resource identifier) from the web server. The requested resource is specified by the URI portion of the GET request.

- POST: The POST method is used to send data to the web server in order to create or modify a resource.

- HEAD: The HEAD method is used to retrieve the HTTP headers that would be returned if a GET method was used for the same URI. The HTTP response corresponding to the HEAD request would not have any content, just HTTP headers.

- PUT: The PUT method creates or replaces the contents of the specified resource on the web server.

- DELETE: The DELETE method deletes the specified resource on the web server.

- CONNECT: The CONNECT method is used to create a tunnel to the server specified in the requested resource.

- OPTIONS: The OPTIONS method is used to retrieve the communication options for the specified resource.

- TRACE: The TRACE method is used mainly for debugging purposes. In the HTTP response, the web server echoes the received request exactly as it was received.

- PATCH: The PATCH method is used to make partial changes to the specified resource.

The GET and POST requests are the two most common HTTP requests. We will look at these a bit closer.

The GET request is used to retrieve data from the server. The GET request does not update the content for the specified resource on the server and, therefore, does not contain a message body. However, the GET request can pass some values via parameters in the URI field, and the server can use the values in the parameter to decide the response.

An example of a GET request is as follows:

```
GET /test/sample.php?name1=value1&name2=value2 HTTP/1.1
Host: mywebserver.com
Connection: keep-alive
User-Agent: Mozilla/5.0 (Windows NT 10.0; Win64; x64)
AppleWebKit/537.36 (KHTML, like Gecko) Chrome/79.0.3945.88
Safari/537.36
Upgrade-Insecure-Requests: 1
Accept: text/html
Accept-Language: en-US,en;q=0.9
Accept-Encoding: gzip, deflate
```

Let's compare the preceding GET request to the HTTP request format given in RFC 2616. The HTTP method, URI, and version are as follows:

- Method: GET

- HTTP URI: /test/sample.php?name1=value1&name2=value2

- HTTP version: 1.1

The remaining section of the HTTP request is the *header* section. Each line in the header section consists of a header option and corresponding value. For example, the User-Agent header option gives details about the application, operating system, and vendor that was used to make the HTTP request. In this case, the User-Agent value is Mozilla/5.0 (Windows NT 10.0; Win64; x64) AppleWebKit/537.36 (KHTML, like Gecko) Chrome/79.0.3945.88 Safari/537.36, which tells us that the request was made from a Chrome browser on a 64-bit Windows 10 device. Similarly, the Host header option lists the web server name or IP address as well as the port to which this request is sent. The Connection header option is a directive to the server whether or not to close the TCP connection after this transaction has been processed. So, `Connection: keep-alive` specifies keeping the TCP connection alive so that more HTTP requests can be made.

HTTP response

Upon receiving the HTTP request, the web server sends a response in return. The response contains a status code that indicates whether the request was successful or not. It also contains a set of HTTP headers, followed by the content or message body.

RFC 2616 specifies the format for the HTTP response as follows:

```
Status-Line
  *(( general-header
  | response-header
  | entity-header ) CRLF)
  CRLF
  [ message-body ]
```

The first line of the response is the status line, which is of the following format:

```
HTTP-Version SP Status-Code SP Reason-Phrase CRLF
```

SP indicates a *space* character. An example of an HTTP response is as follows:

```
HTTP/1.1 200 OK
Date: Thu, 13 May 2004 10:17:12 GMT
Server: Apache
Last-Modified: Tue, 20 Apr 2004 13:17:00 GMT
ETag: "9a01a-4696-7e354b00"
Accept-Ranges: bytes
Content-Length: 18070
Keep-Alive: timeout=15, max=100
Connection: Keep-Alive
Content-Type: text/html; charset=ISO-8859-1

<?xml version="1.0" encoding="UTF-8"?>
<!DOCTYPE html
  PUBLIC "-//W3C//DTD XHTML 1.0 Strict//EN"
  "DTD/xhtml1-strict.dtd">
<html xmlns="http://www.w3.org/1999/xhtml" xml:lang="en" lang="en">
  <head>
    <title>Ethereal: Download</title>
```

```
<style type="text/css" media="all">
 @import url("mm/css/ethereal-3-0.css");
</style>
```

In the preceding example, the status code is 200 and the status message is OK. This is the most common status code and message. HTTP server status codes are grouped into various categories, as follows:

- 1xx: Informational

- 2xx: Successful

- 3xx: Redirection

- 4xx: Client error

- 5xx: Server error

Based on the range the status code falls under, we can establish the status of the request – whether it was successful, whether it resulted in an error, and so on. As we saw in the example, the status code of 200 meant that the request was successful. A few other status codes and their corresponding reason codes (status messages) are given here:

- 100 – Continue:

  This is a status code used by the server to communicate to the client that it can proceed with the request and complete it.

- 301 – Moved Permanently:

  This is the response from the web server when the requested resource has been moved or relocated:

  ```
  HTTP/1.1 301 Moved Permanently
  Location: http://www.newlocation.org/index.asp
  ```

  In such cases, the response header will contain a Location header option with the new URI that contains the requested resource, as shown in the preceding code.

- 404 – Not Found:

  This is the status code that the server sends when the requested resource is not present or not found. This is typically seen when someone types in the wrong URL into the browser. For example, if a user types in https://recsports.uga.edu/aboutus/ instead of https://recsports.uga.edu/about-us/, it results in a 404 status code (notice the missing hyphen in the URL).

- `503 - Service Unavailable:`

  This is a common server error-related status code that is seen when the server is unable to provide a reply due to temporary overload.

  Right after the first line starts, the header section for the HTTP response continues till the double **Carriage Return-Line Feed (CRLF)** pattern.

Let's briefly take a look at the various header options in the HTTP response.

The `Server` header option describes the web server that provided the HTTP response. This option is not encouraged as it reveals information about the server to a potential malicious end user. The `Last-Modified` header option provides the time and date when the resource requested was last modified (according to the responding web server). The `Etag` option is an identifier that indicates a specific version of the resource provided by the HTTP response. This is more accurate and commonly used than the `Last-Modified` option. The `Content-Length` option provides the size of the HTTP response data.

The HTTP header options end and are followed by the double CRLF pattern, and then the HTTP response data is provided.

HTTP/1.1 was a text-based protocol and was widely adopted and used. In the next section, let's look at HTTP/2, which is the successor of HTTP/1.1.

HTTP/2

HTTP/2 is the next major version of HTTP after HTTP/1.1. This version is being increasingly supported and used on web browsers and web servers. The main aim of HTTP/2 was to improve the efficiency of HTTP.

Some of the key differences between HTTP/2 and HTTP/1.1 include the following:

- **Binary frames**: HTTP/2 uses a binary frame structure. On the other hand, HTTP/1.1 was a text-based protocol. Each of the HTTP/2 frames has the following fields: *Length*, *Type*, *Flags*, *Stream Identifier*, and *Frame Payload*.

- **Multiplexing**: Once the HTTP/2 connection is established, the client and server communicate using multiple and separate streams. A stream is a logical channel between the client and server through which the HTTP frames are sent. Each stream has a unique stream identifier that is specified in the frame.

- **Server Push**: This is yet another key feature in HTTP/2, which is again geared toward improving efficiency. The Server Push feature enables the HTTP web server to send multiple responses to a single HTTP request from the client. The server can anticipate a future request from the client and proactively *push* the response before the client makes a request.

In short, the main difference in HTTP/2 is how the data is represented and sent over the network, how the requests and responses are encoded and structured, and how simultaneous requests/responses are made possible.

In this section, we briefly discussed the basics of HTTP. Let us take a look at the Snort HTTP inspector modules next.

HTTP inspector

Snort 3 supports two HTTP inspectors: `http_inspect` and `http2_inspect`. We will look at both inspectors in this section.

The HTTP Inspect (inspector) module analyzes the HTTP requests and responses. The module depends on the Stream TCP inspector to provide TCP data as a continuous (reassembled) stream of bytes. The module has a splitter component – `HTTPStreamSplitter` – that divides the TCP stream into HTTP **protocol data units** (**PDUs**). The HTTP PDUs are individual HTTP requests and responses. The operation of the HTTP Inspector module is shown at a high level in *Figure 10.1*:

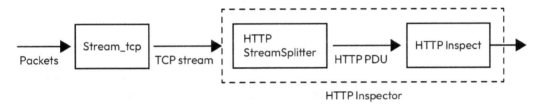

Figure 10.1 – The stream_tcp module provides a stream of bytes to
the HTTP inspector, which divides into HTTP PDUs

The HTTP inspector analyzes, parses, and makes available the HTTP message in sections. Subsequently, the following message sections of an HTTP request are available to be matched against for detection:

- **Request line**: This is the first line of an HTTP request, as described in the previous section

- **Header section**: This is the HTTP request header section that starts on the next line after the request line and goes on till the double **Carriage Return-Line Feed** (**CRLF**) pattern

- **Message body**: This pertains to a `POST` request, and will be associated with a `Content-Length` header option

- **Old message body**: This pertains to a message body that is sent without a corresponding `Content-Length` header option

Similarly, the following message sections of an HTTP response are available to be matched against for detection:

- **Status line**: This is the first line of an HTTP response, as described in the previous section.

- **Headers**: This is the entire HTTP header section that is present in the HTTP response message, which starts right after the first line in the HTTP response and goes on till the double CRLF pattern.

- **Response message body**: This is the content that follows the header section. This section contains the resource or data that was requested using the HTTP request.

- **Chunked message body**: In certain cases, the requested content size may not be known when the server starts sending the data to the client. The data is sent in *chunks* or data blocks. Each chunk is preceded by the chunk size. The last chunk will be of zero size.

We saw how the HTTP request and response messages have various components such as the method used (GET, POST, etc.), the URI requested, the various header options, the HTTP response message, and so on. In order to perform accurate and contextual analysis, Snort parses out these components into specific buffers, which are then available for inspection during the signature matching phase. In the next section, we will look at some of these buffers, which are also known as HTTP buffers.

HTTP buffers

When the HTTP inspector analyzes the HTTP message (namely, the *request* and the *response*), it parses out certain key elements into specific buffers. This enables more precise matching and reduces false positives. For example, if an OS command injection is applicable for a certain parameter in the HTTP URI, it will be more accurate to detect it against the URI alone rather than the entire HTTP request.

In Snort 3, when referenced in a rule, these buffers are called **sticky buffers**. This means that once the buffer is specified, all the subsequent matching is specific to that buffer until the stickiness is altered. (In Snort 2, the sticky aspect was not there; therefore, each content match or **Perl Compatible Regular Expression** (**PCRE**) had to be qualified with the name of the buffer (or a corresponding option for PCRE). We will discuss the PCRE and content match options in *Chapter 13*.

The various buffers that are available and populated by the inspector are as follows. The name of the buffer as referred to in a signature is given in parentheses:

- **HTTP request method** (http_method): This buffer contains the name of the HTTP request method used (GET, POST, etc.).

- **HTTP URI** (http_uri): The entire URI is parsed and populated into this buffer. This contains the normalized URI.

- **HTTP raw URI** (http_raw_uri): The entire URI is parsed and populated into this buffer as it was seen on the wire. This buffer contains the URI without any normalizations.

- **HTTP version** (`http_version`): This contains the version string that is present in the HTTP request line.

- **HTTP param** (`http_param`): The various query parameters and values that are present in the HTTP URI are present in this buffer.

- **HTTP raw request** (`http_raw_request`): The entire HTTP request line (unnormalized) is copied into this buffer.

- **HTTP header** (`http_header`): The entire HTTP header is copied into this buffer. The header options are in the same order as in the original request. The appropriate header normalizations are done to the content in this buffer.

- **HTTP raw header** (`http_raw_header`): This buffer contains non-normalized HTTP header bytes.

- **HTTP cookie** (`http_cookie`): This buffer contains the normalized cookie string.

- **HTTP raw cookie** (`http_raw_cookie`): This buffer contains the original cookie string.

- **HTTP client body** (`http_client_body`): This is the normalized message section of a POST request.

- **HTTP raw client body** (`http_raw_body`): This is the unnormalized message body of a POST request. The Content-Length value in the request matches the length of this section.

- **HTTP status code** (`http_stat_code`): The HTTP response status code is extracted into this buffer.

- **HTTP status message** (`http_stat_message`): The HTTP response status message is extracted into this buffer.

- **HTTP raw status** (`http_raw_status`): The unnormalized HTTP status line is copied into this buffer.

- **HTTP trailer** (`http_trailer`): The normalized HTTP trailer string is copied to this buffer.

- **HTTP raw trailer** (`http_raw_trailer`): This buffer contains the unnormalized HTTP trailer.

- **True IP** (`http_true_ip`): The original IP address of the client is extracted from the appropriate HTTP header into this buffer.

These buffers and the corresponding keywords play a key role in writing HTTP-related Snort signatures. We will see more of this in *Chapter 13*.

The next HTTP inspector supported by Snort is the HTTP/2 inspector, which we will discuss in the next section.

HTTP/2 inspector

The approach Snort takes is as follows. The `http2_inspect` inspector essentially parses and analyzes the HTTP/2 message frames and converts/translates them into an HTTP/1.1 message. Thereafter, these messages are sent through the HTTP Inspect inspector.

The `Http2StreamSplitter` component of the `http2_inspect` module parses the HTTP/2 traffic and separates/splits the HTTP/2 frames. Thereafter, these are translated into HTTP/1.1 messages so that they can be processed and matched against the signatures.

In the next section, let us discuss the various configuration settings for the HTTP Inspector modules.

HTTP inspector configuration

The HTTP inspectors are enabled via the configuration file as follows:

```
http_inspect = { }
http2_inspect = { }
```

These enable the corresponding inspectors to use the default configuration settings. Any deviations from the default settings need to be done by modifying the specific configuration parameters. Each of the inspectors has a set of their own configurable parameters.

The configuration options for each module can be listed using the `help` command available with Snort. For example, the command for `http_inspect` would be as follows:

```
snort3 --help-module http_inspect
```

Let us briefly discuss a few of the configuration options available for `http_inspect`:

- **Limit to request and response depth**: The request and response sizes can be really large. Therefore, in order to limit the inspection for performance reasons, there are two configuration settings, as follows:

  ```
  int http_inspect.request_depth = -1: max request message body
  bytes to examine
  int http_inspect.response_depth = -1: max response message body
  bytes to examine
  ```

- **Maximum length of an HTTP chunk**: HTTP chunked encoding is a technique used by the HTTP/1.1 version when the length of the content is not known before starting to send the data. In such cases, the transfer is done using an encoding called **chunked encoding**. In this technique, the data to be sent is divided into non-overlapping chunks. Each chunk is preceded by its length in bytes. The length of the chunk is specified as a hexadecimal value.

  This configuration parameter sets a maximum allowed value for an HTTP chunk:

  ```
  http_inspect.maximum_chunk_length= 4294967295
  ```

- **Maximum header length and maximum header options**: HTTP does not limit the size of the HTTP header. However, various servers put a limit on the header length. For example, Apache sets a limit of 8,192 bytes for the HTTP header.

  Snort, by default, sets a limit of 4,096 bytes for the maximum HTTP header length. This can be configured using the following parameter:

  ```
  int http_inspect.maximum_header_length = 4096
  ```

  Similarly, Snort also has a parameter that limits the number of HTTP header options and values specified in a PDU:

  ```
  int http_inspect.maximum_headers = 200
  ```

  HTTP/1.1 has a pipelining feature, where a client (web browser) can send multiple HTTP requests to the server on separate TCP connections. Such requests are typically related to each other. Snort has a parameter that limits the number of such pipelined requests, as follows:

  ```
  int http_inspect.maximum_pipelined_requests = 99: alert when the
  number of pipelined requests exceeds this value { 0:99 }
  ```

- **True client IP header options**: This configuration parameter specifies the order of priority of the HTTP header options that must be parsed to populate the true IP buffer. Typically, the true client IP address is contained within the x-forwarded-for and true-client-ip header. If the header section contains both these options, Snort 3 needs to know which one should be prioritized over the other.

  For example, http_inspect.xff_headers = 'x-forwarded-for true-client-ip' specifies that x-forwarded-for is the first priority followed by true-client-ip:

  ```
  string http_inspect.xff_headers = 'x-forwarded-for true-
  client-ip'
  ```

The command for http2_inspect would be as follows:

```
snort3 --help-module http2_inspect
```

The http2_inspect inspector only has a single configuration parameter, as we see from the output of the aforementioned command:

```
Configuration:
int http2_inspect.concurrent_streams_limit = 100: Maximum number of
concurrent streams allowed in a single HTTP/2 flow { 100:1000 }
```

This parameter (concurrent_streams_limit) specifies a higher ceiling for the number of streams that are allowed in a single HTTP/2 connection. When this limit is exceeded, the module generates an alert:

```
(http2_inspect) excessive concurrent HTTP/2 streams".
```

The HTTP configuration in Snort is very rich, and we have covered a portion of it in this section. However, this gives you the needed familiarity with how to go about configuring the HTTP inspector.

Summary

In this chapter, we looked at the HTTP inspector, which is one of the key modules of Snort 3. We briefly discussed HTTP and also looked at the function of the inspector. We built up the necessary background to understand how the various HTTP-related keywords would work from a Snort rule-writing perspective. Finally, we looked briefly at a few configuration parameters.

In the next chapter, we will discuss the SMB and DCE RPC protocols, their usage, and how the DCERPC inspector analyzes these protocols.

11
DCE/RPC Inspectors

A **Remote Procedure Call** (**RPC**) enables programs on one system (computer) to invoke a procedure or function on a different system (computer). An RPC is a network protocol that the calling program (client) uses to communicate with the service on a system (server) that implements the called procedure. DCE/RPC is a protocol that enables the RPC feature in a **Distributed Computing Environment** (**DCE**) system. Microsoft's implementation of the DCE/RPC protocol is referred to as **MSRPC**.

In recent years, the DCE/RPC protocol has been used as an attack vector by bad actors. Some of the recently noted DCE/RPC-related vulnerabilities include the PrintNightmare vulnerabilities (CVE-2021-1675, CVE-2021-34527, and CVE-2021-34527), the zero-click exploit targeting Microsoft RPC services (CVE-2022-26809), and the out-of-bounds write vulnerability on vCenter Server (CVE-2023-34048).

Detecting the exploits and attacks against the DCE/RPC protocol is important. The DCE/RPC inspectors on Snort 3 analyze the DCE/RPC protocol as seen on the network and enable the detection of DCE/RPC-based attacks.

In this chapter, we will take an overview of DCE/RPC, as well as discuss its inspectors, dependencies, relevant rule options, and configurations:

- A DCE/RPC overview
- DCE/RPC inspectors
- DCE/RPC rule options
- Exercise

A DCE/RPC overview

An RPC is a mechanism in distributed computing where a computer program (the caller procedure) executing in one context can invoke a procedure executing in a different context (the called procedure), using the help of a network communication protocol. In an RPC, the entity that makes the procedure invocation is referred to as the RPC client, and the entity that receives and executes the procedure is the RPC server. To make the procedure call, the RPC client must establish a connection to the server, which is referred to as binding.

According to www.dcerpc.org, DCE/RPC is an implementation of the RPC technology developed by the Open Group as part of the Distributed Computing Environment. DCE/RPC is mostly used to interact with Windows network services.

RPC interfaces are a set of procedures that a server offers and that a client can invoke. An RPC server can have any number of interfaces. The RPC interfaces have a unique identifier called an **interface UUID**, which uniquely identifies the RPC interface that is called. The interfaces offered by the RPC server also have major and minor versions associated with them.

The DCE/RPC network protocol is the communication protocol that enables the DCE/RPC mechanism to operate as mentioned. The DCE/RPC protocol is designed to operate over both connection-oriented **Server Message Block (SMB)** and **Transmission Control Protocol (TCP)** and connectionless layers.

In the DCE/RPC protocol, an endpoint refers to a network-specific address that will be used to make remote procedure calls. This is typically a port (TCP or UDP) or a named pipe name in the case of SMB. An endpoint can be a well-known endpoint or a dynamic endpoint.

An endpoint mapper is a service on the DCE/RPC server that maintains a database that maps the interfaces supported by the DCE/RPC server to the endpoint address. Servers add/delete information to/from this database by registering/unregistering RPC interfaces. Usually, the endpoint mapper service listens on TCP port 135 or the named pipe – \\PIPE\epmapper. (You can read more about named pipes at https://learn.microsoft.com/en-us/windows/win32/ipc/named-pipes.)

Let's look at an example of the DCE/RPC traffic related to the endpoint mapper (Higher resolution images for these diagrams are available at github.com/PacktPublishing/IDS-and-IPS-with-Snort-3.0/ under the *Chapter 11* folder).:

No.	Time	Source	Destination	Protocol	Length	Info
1	0.000000	10.0.2.15	192.168.3.43	TCP	66	51285 → 135 [SYN] Seq=0 Win=8192 Len=0 MSS=1460 WS=256 SACK_PERM
2	0.000687	192.168.3.43	10.0.2.15	TCP	60	135 → 51285 [SYN, ACK] Seq=0 Ack=1 Win=65535 Len=0 MSS=1460
3	0.000771	10.0.2.15	192.168.3.43	TCP	54	51285 → 135 [ACK] Seq=1 Ack=1 Win=64240 Len=0
4	0.001010	10.0.2.15	192.168.3.43	DCERPC	214	Bind: call_id: 2, Fragment: Single, 3 context items: EPMv4 V3.0 (32bit NDR), EPMv4 V3.0 (64bit NDR...
5	0.001147	192.168.3.43	10.0.2.15	TCP	60	135 → 51285 [ACK] Seq=1 Ack=161 Win=65535 Len=0
6	0.018774	192.168.3.43	10.0.2.15	DCERPC	162	Bind_ack: call_id: 2, Fragment: Single, max_xmit: 5840 max_recv: 5840, 3 results: Provider rejecti...
7	0.018974	10.0.2.15	192.168.3.43	EPM	222	Map request, WITNESS, 32bit NDR
8	0.019225	192.168.3.43	10.0.2.15	TCP	60	135 → 51285 [ACK] Seq=109 Ack=329 Win=65535 Len=0
9	0.020961	192.168.3.43	10.0.2.15	EPM	226	Map response, WITNESS, 32bit NDR
10	0.068276	10.0.2.15	192.168.3.43	TCP	54	51285 → 135 [ACK] Seq=329 Ack=281 Win=63960 Len=0
11	15.010756	10.0.2.15	192.168.3.43	TCP	54	51285 → 135 [FIN, ACK] Seq=329 Ack=281 Win=63960 Len=0
12	15.011115	192.168.3.43	10.0.2.15	TCP	60	135 → 51285 [ACK] Seq=281 Ack=330 Win=65535 Len=0
13	15.012027	192.168.3.43	10.0.2.15	TCP	60	135 → 51285 [FIN, ACK] Seq=281 Ack=330 Win=65535 Len=0
14	15.012092	10.0.2.15	192.168.3.43	TCP	54	51285 → 135 [ACK] Seq=330 Ack=282 Win=63960 Len=0

Figure 11.1 – The Wireshark display shows the DCE/RPC endpoint mapper-related traffic on port 135

This is the connection-oriented DCE/RPC. We can see that the TCP connection is established by the first three packets (a TCP three-way handshake), followed by the DCE/RPC binding sequence using the bind request and response. Subsequently, the query is made to the endpoint mapper service using the Map-Request packet. We will look into the Map-Request and Map-Response packets in detail in *Figure 11.2* and *Figure 11.3* respectively.

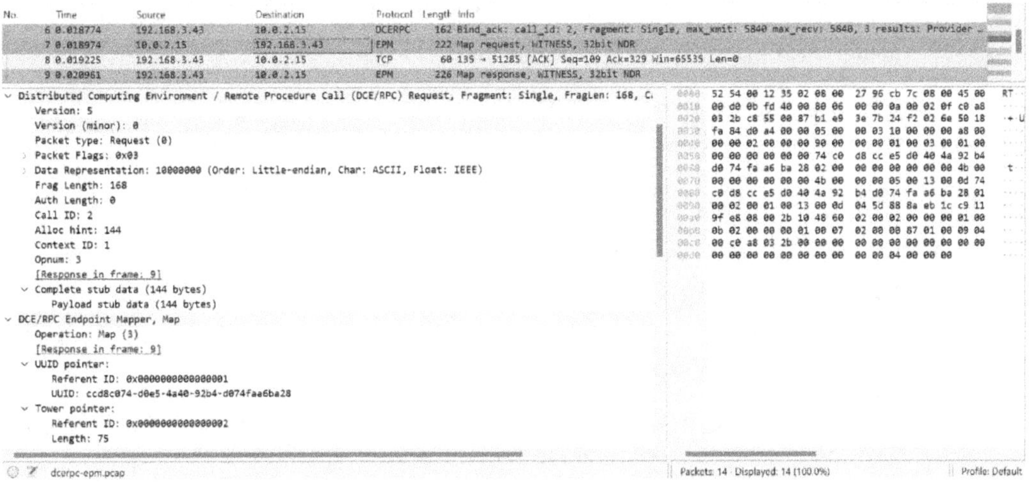

Figure 11.2 – The Wireshark display focuses on the DCE/RPC EPM Map-Request packet

The DCE/RPC Map-Request packet specifies the UUID of the service it needs the endpoint address to – namely, Witness (an example service). Once the server processes this request, it responds by sending the Map-Response packet with the answer.

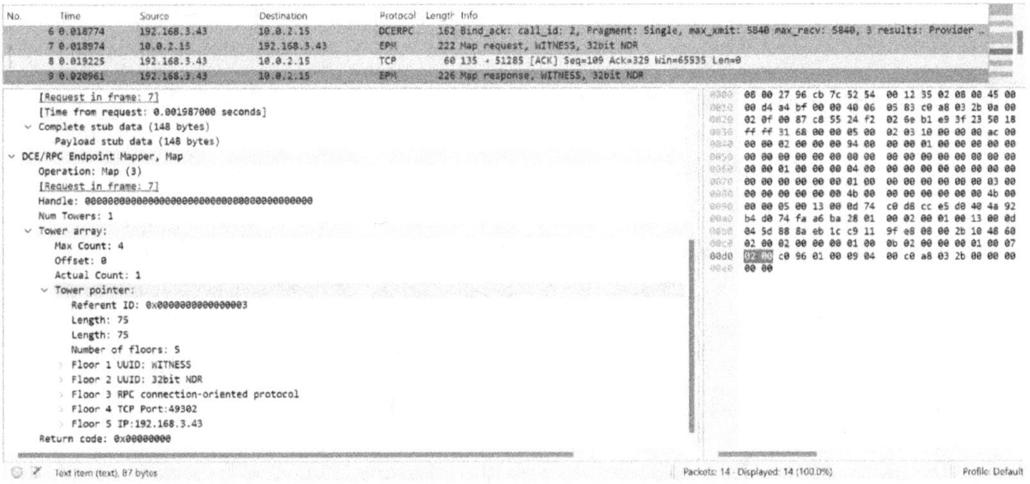

Figure 11.3 – The Wireshark display focuses on the DCE/RPC EPM Map-Response packet

The Map-Response packet specifies the IP address and TCP port – the address for the dynamic endpoint.

In the next section, let's look at the different types of DCE/RPC.

Connectionless versus connection-oriented DCE/RPC

A connection-oriented DCE/RPC depends on a transport that ensures that there is a connection between the client and server, and that the delivery is reliable. For example, TCP is a protocol that is connection-oriented, ensuring reliable data transmission.

A connectionless DCE/RPC uses a transport that is not connection-oriented or, in other words, connectionless. A typical example of a connectionless protocol is UDP.

There are four different types of transports that DCE/RPC uses:

- **DCE/RPC over TCP**: This case is also referred to as `ncacn_ip_tcp`. In this case, the DCE/RPC packets are sent over the TCP protocol.
- **DCE/RPC over UDP**: This case is referred to as `ncadg_ip_udp`. The DCE/RPC packets will be sent directly over UDP.
- **DCE/RPC over SMB**: In this case, the DCE/RPC packets are sent over the **Server Message Block (SMB)** protocol. This case is referred to as the `ncacn_np` transport.
- **DCE/RPC over HTTP**: In this case, the DCE/RPC packets are sent over TCP after the setup is done, using HTTP packets.

In the next section, we will discuss the different DCE/RPC inspectors based on the differences in the transport protocol used.

DCE/RPC inspectors

The DCE/RPC inspectors inspect and analyze the various types of DCE/RPC traffic that use the following transports – namely, TCP, UDP, SMB, and HTTP. This covers both connection-oriented as well as connectionless DCE/RPC cases.

In this section, we will discuss briefly how the DCE/RPC inspectors process the incoming traffic. We will also discuss the configuration of different DCE/RPC inspectors.

The DCE/RPC inspector traffic analysis ensures the following points:

- Detects anomalous and evasion attempts using DCE/RPC protocol characteristics.
- Enables the rules engine to match against DCE/RPC traffic using protocol-specific rule options.

Based on the transport protocol that is used, we have the following DCE/RPC inspectors – namely, `dce_tcp`, `dce_udp`, `dce_smb`, and `dce_http`.

Of these, `dce_udp` is the only connectionless protocol. All the others are connection-oriented. In the case of DCE/RPC over HTTP, only the setup phase happens over HTTP; once the setup is over, the protocol is similar to DCE/RPC over TCP (dce_tcp). When it comes to processing, there is some commonality among the connection-oriented protocols.

An incoming session is assigned to the right service inspector by the Binder and Wizard inspector modules. The Binder module uses a set of specified rules to assign an incoming session to a particular service inspector, whereas Wizard uses a combination of special content to determine what protocol is involved, and then it assigns it to the relevant service inspector. In the case of DCE/RPC over TCP, SMB, and HTTP, the processing involves connection-oriented processing.

The connection-oriented processing handles transport layer segmentation and uses a buffering approach to gather data till it receives a complete **Protocol Data Unit** (**PDU**). Once the complete PDU is obtained, the decoding of the PDU is performed.

At the same time, each packet may contain more than one PDU. In such cases, the processing loops through each PDU and sends it for decoding. Among the various PDU types, the following are currently recognized and analyzed – **bind**, **alter context**, and **request**.

During the decoding phase, information such as **interface**, **opnum**, and **stub data** is parsed and/or decoded such that signatures can be applied against them. After the decoding is done, the PDU is sent through the detection engine to match against signatures.

Let's look at the different DCE/RPC inspectors and the relevant configurations next:

- dce_tcp: The dce_tcp inspector is the inspector module that analyzes DCE/RPC over TCP transport. The DCE/RPC protocol works directly on top of the TCP protocol layer in this case. DCE/RPC over TCP is a connection-oriented option.

  The following table shows the configuration parameters for the dce_tcp inspector:

Config parameter	Usage
limit_alerts	If set to true, this parameter limits the number of DCE/RPC alerts to, at most, one per signature per flow. The default value is true.
disable_defrag	If set to true, this parameter disables the defragmentation feature. The default setting is false.
reassemble_threshold	This parameter specifies the number of bytes that will be buffered before triggering a reassembly operation, sending the reassembled buffer through the rules engine.

Config parameter	Usage
policy	This parameter specifies the version for Windows or Samba, such as WinXP or Win2003, or Samba-3.0.20. The configuration could have several policies, based on the version of Windows or Samba identified for the session.
max_frag_len	This parameter specifies the maximum allowed fragment length for defragmentation.

Table 11.1 – Configuration parameters for dce_tcp module

Next, let's look at the dce_udp inspector module:

- dce_udp: The dce_tcp inspector is the inspector module that analyzes DCE/RPC over UDP transport. This is the connectionless transport case.

  The following table shows the configuration parameters for the dce_udp inspector:

Config parameter	Usage
limit_alerts	If set to true, this parameter limits the number of DCE/RPC alerts to, at most, one per signature per flow. The default value is true.
disable_defrag	If set to true, this parameter disables the defragmentation feature. The default setting is false.
max_frag_len	This parameter specifies the maximum allowed fragment length for defragmentation.

Table 11.2 – Configuration parameters for dce_udp module

Next, let's look at the dce_smb inspector module:

- dce_smb: This is the DCE/RPC inspector for DCE/RPC over SMB transport. The inspector covers both SMBv1 and SMBv2. The inspector also performs SMB desegmentation to avoid evasion attempts. This is a connection-oriented service, where DCE/RPC uses authenticated named pipes over the SMB or SMB2 protocol.

The -help-module dce_smb command option can be used to list the configuration parameters related to the dce_smb inspector. The following table shows the configuration parameters for the dce_smb inspector:

Config parameter	Usage
limit_alerts	If set to true, this parameter limits the number of DCE/RPC alerts to, at most, one per signature per flow. The default value is true.
disable_defrag	If set to true, this parameter disables the defragmentation feature. The default setting is false.
reassemble_threshold	This parameter specifies the number of bytes that will be buffered before triggering a reassembly operation, sending the reassembled buffer through the rules engine.
smb_fingerprint_policy	This parameter specifies what traffic to use to fingerprint the Windows or Samba version. The options are none, client, server, or both. The default is none. The fingerprint identification value is used to select which policy to use.
policy	This parameter specifies the version for Windows or Samba, such as WinXP or Win2003, or Samba-3.0.20. The configuration could have several policies, based on the version of Windows or Samba identified for the session.

Config parameter	Usage
smb_max_chain	Specifies the maximum allowed value for the number of chained SMB AndX commands. If the traffic contains more chained SMB AndX commands than the specified value for this parameter, an alert is triggered.
smb_max_compound	The maximum values for allowed commands in a single SMB request.
valid_smb_versions	Sets the versions of SMB that should be analyzed.
smb_file_inspection	This parameter enables or disables the SMB file inspection feature.
smb_file_depth	Sets the amount of bytes of the file transferred via SMB that will be inspected.
smb_invalid_shares	Specifies a list of SMB-shared resources that will be monitored and alerted if encountered – for example, C$.
memcap	Sets a limit to the memory usage for the inspector. When this limit is exceeded, an alert/log is generated.

Table 11.3 – Configuration parameters for dce_smb module

Next, let's look at the dce_http inspector module:

- dce_http: The dce_http inspector is the inspector module that analyzes DCE/RPC over HTTP transport. This is the connection-oriented transport case. In this case, the DCE/RPC protocol works on top of the TCP protocol; the only difference is that there is an additional setup phase that happens over the HTTP protocol.

There are no configuration parameters for this inspector.

In this section, we briefly looked at the different DCE/RPC inspectors and their relevant configuration parameters. In the next section, we'll discuss the different rule options that are connected to the DCE/RPC feature.

DCE/RPC rule options

There are three Snort rule options that are provided by the DCE/RPC inspector functionality:

- `dce_iface`: The DCE/RPC interfaces have a unique identifier called interface UUID that uniquely identifies the DCE/RPC interface that is called. The interfaces advertised by the server also have major and minor versions associated with them.

  This rule option takes in UUID as a parameter, a version, and frag settings.

  If the UUID, version, and frag settings specified in the rule match the DCE/RPC request that is seen on the wire, the rule option succeeds.

- `dce_opnum`: For a DCE/RPC request, opnum represents a specific function for the interface that is called. This rule option takes a number, a list of numbers, or a range.

  If the operation number of the DCE/RPC request seen in the traffic matches the specified list or range of numbers of the rule option, it is considered a successful match for the rule option.

- `dce_stub_data`: This rule option is used to detect whether there is DCE/RPC stub data in the analyzed traffic.

These are the rule options relevant to the DCE/RPC feature, and the options that depend on the DCE/RPC inspectors.

In the next section, we will recap all that we have learned in this chapter by doing an exercise.

Exercise

Let's do an exercise that encapsulates all the things we learned together. We will use the packet capture file we discussed in this chapter that has the DCE/RPC traffic to the Witness interface. This packet capture can be downloaded at https://wiki.wireshark.org/uploads/__moin_import__/attachments/SampleCaptures/dcerpc_witness.pcapng

We will make sure that the DCE/RPC inspectors are enabled in the `lua` configuration file:

```
dce_smb = { }
dce_tcp = { }
dce_udp = { }
dce_http_proxy = { }
dce_http_server = { }
```

We will also create a Snort signature that will detect DCE/RPC traffic that uses the Witness interface (UUID `ccd8c074-d0e5-4a40-92b4-d074faa6ba28`):

```
alert tcp any any -> any any (msg:"DCERPC Witness Interface";
flow:established; dce_iface: uuid ccd8c074-d0e5-4a40-92b4-
d074faa6ba28; sid:2344; priority:3;)
```

We will run the `snort` command as follows:

```
./build/src/snort -c lua/snort.lua -R ~/Rules/local.rules -r ~/dcerpc_
witness.pcapng -k none -l ~/Log
```

We will observe the generated alerts in the `~/Log` directory:

```
cat ~/Log/alert_fast.txt
09/23-05:05:48.122542 [**] [1:2344:0] "DCERPC Witness Interface" [**]
[Priority: 3] {TCP} 10.0.2.15:51296 -> 192.168.3.43:49302
09/23-05:05:48.122865 [**] [1:2344:0] "DCERPC Witness Interface" [**]
[Priority: 3] {TCP} 192.168.3.43:49302 -> 10.0.2.15:51296
```

There are other DCE/RPC-related packet captures available at `https://wiki.wireshark.org/SampleCaptures` if you wish to run additional exercises.

In this section, we saw how to enable the DCE/RPC inspector using configuration, as well as how to craft a rule that uses DCE/RPC-related rule options to detect something specific. The resultant Snort alert shows the successful rule match.

Summary

Since DCE/RPC is increasingly becoming an attack vector, the role of DCE/RPC analysis is a key component of Snort. In this chapter, we discussed the DCE/RPC protocol and the Snort inspectors related to DCE/RPC. We discussed briefly the workings of the protocol, as well as connection-oriented and connectionless DCE/RPC. We also covered the different inspectors as well as the related configuration.

In the next chapter, we will discuss the topic of IP reputation in an IDS and the IP reputation inspector in Snort 3.

12
IP Reputation

The concept of reputation is used commonly in our daily lives. It is often used as a metric to evaluate an entity, be it a place, a person, a group or an organization, a school or a company, and so on. The reputation associated with a particular entity is made and/or adjusted based on past and ongoing experiences, whether good or bad. A good reputation will open doors to incredible opportunities, whereas a bad reputation will result in closed doors. IP reputation follows the same concept. The object with which we associate the reputation is an IP address.

In this chapter, we will look at the IP reputation inspector module in Snort 3. We will discuss the following topics:

- Background
- Configuration of the IP reputation inspector module
- Functionality of the IP reputation inspector
- IP reputation inspector – alerts and pegs

Background

Attribution is one of the steps during an incident response activity following a cyber-attack or malicious activity. One aspect of this step is IP address attribution, namely identifying the source IP address(es) that was responsible for the attack.

IP blacklisting is a process where we maintain a list of IP addresses that are repeatedly involved in malicious activity. Individual organizations can maintain their own IP blacklists and share this information with other organizations in order to create a more comprehensive and reliable IP blacklist.

These lists are then used on firewalls, routers, intrusion prevention systems, and/or individual machines in order to block any traffic to and from any IP address that is present in it.

The IP blacklisting process typically involves comparing the source and destination IP address of every packet against the list of IP addresses on the blacklist. When the blacklist contains several thousands of IP addresses, this comparison process can become computationally very expensive if not done efficiently.

IP address as an entity – use of blocklists

Public IP addresses are often considered globally unique at any point in time. This means that at any point in time, a public IP address is mapped to a unique device that is connected to the internet. This is not true for the case of private IP addresses, since that range of IP addresses is available to be reused within private networks. Since public IP addresses are considered unique globally, it is treated as an entity. In other words, since public IP addresses are uniquely attributed to a certain machine or device on the internet, that address can be reliably used as an identifier.

Based on this condition, historically, the security community started creating lists of *bad* IP addresses. For example, the DShield project started publishing the top 10 source IP addresses that were involved in attacks. Later, this idea developed into various groups maintaining and publishing IP blocklists based on the attack data from collaborating organizations.

Challenges

There are a few challenges to the IP blacklisting process:

- **Shared or temporary IP addresses**: A user that is connected to the internet via an **Internet Service Provider (ISP)** gets a public IP address. The protocol used in such cases for IP address assignment is the **Dynamic Host Configuration Protocol (DHCP)**. The public IP that is assigned is globally unique at any point in time, as we have discussed. However, it is not globally unique across an extended period. For example, User 1 may get a particular IP when connected to the ISP; they use the IP for a time and then get disconnected. After a few days, that same IP may be reassigned to User 2. So, if the IP address is added to a blocklist due to the malicious activity of User 1, it will affect User 2, who may be an innocent user.

- **IP addresses can be spoofed**: It is pretty straightforward to spoof an IP address, that is, to send a packet with a forged source IP address. Nowadays, there are various egress filtering systems that prevent this. In addition, successfully making a TCP connection with a spoofed IP address is not trivial, but it is still doable.

- **Network Address Translation (NAT)**: NAT is used when there are several devices all using the same public IP address. The NAT operation is usually done by a firewall or router. When NAT is present, the attribution cannot be done to a single device, but only to one of the devices in the NAT-ed network.

- **Proxy server**: There are proxy servers that proxy various services, such as HTTP proxy. When the traffic is coming through the proxy, the source IP address will be the address of the proxy. Hence, the IP attribution will not work as needed.

Even though we have these challenges, IP blacklisting is still a very popular technique used in network and endpoint security. In the next section, we will discuss the history of IP blacklisting using Snort.

History of IP blocking in Snort

Historically, Snort (before version 2.9.1) in IPS mode implemented IP blocking using the traditional Snort drop rules. Let's look at an example:

```
drop ip 85.145.131.51 any -> $HOME_NET any (msg:"Drop Traffic From
85.145.131.51"; sid:1234567; priority:1; classtype:bad-unknown; )
```

This is a drop rule that drops all IP datagrams with 85.145.131.51 as the source IP address and $HOME_NET matching the destination IP address. In a realistic situation, the number of IP addresses to be blocked is in the hundreds or thousands, and the number of such rules is high. Each of these rules will have to be evaluated one after another in a linear fashion and for every single packet. This affects the performance of the Snort IPS.

With Snort 2.9.1, the IP reputation module was implemented as a preprocessor, and it provided the capability of blocking a set of IP addresses. This solved the problem of managing the IP blocklisting and performance issues. With the advent of Snort 3, the functionality was ported over as an inspector.

> **Terminology clarification**
>
> The terms *blacklist* and *blocklist* are used interchangeably. In the context of Snort, they mean the same thing. When we discuss the IP reputation configuration and functionality, we will see these terms being used interchangeably. Similarly, the terms *whitelist* and *allowlist* are also used interchangeably.

We will discuss the IP reputation inspector module in the next section.

Configuration of the IP reputation inspector module

In this section, we will briefly discuss the various configuration parameters associated with the IP reputation inspector. The --help-config command line option is useful for retrieving configuration information for any module. Here, we can see a list of configuration parameters that are relevant to the IP reputation inspector module:

```
$ snort3 --help-config reputation
string reputation.blocklist: blocklist file name with IP lists
string reputation.list_dir: directory for IP lists and manifest file
int reputation.memcap = 500: maximum total MB of memory allocated {
1:4095 }
enum reputation.nested_ip = 'inner': IP to use when there is IP
encapsulation { 'inner|outer|all' }
enum reputation.priority = 'allowlist': defines priority when there is
```

```
a decision conflict during run-time { 'blocklist|allowlist' }
bool reputation.scan_local = false: inspect local address defined in
RFC 1918
enum reputation.allow = 'do_not_block': specify the meaning of
allowlist { 'do_not_block|trust' }
string reputation.allowlist: allowlist file name with IP lists
```

In addition, there are two variables that are relevant to the inspector present in the snort_defaults. lua configuration file. These are WHITE_LIST_PATH and BLACK_LIST_PATH; they specify the directory location of the whitelist and blacklist files:

```
-- If you are using reputation preprocessor set these
WHITE_LIST_PATH = '../lists'
BLACK_LIST_PATH = '../lists'
```

Let's briefly look at the various configuration parameters of the IP reputation inspector module that we listed using the --help-config command:

- allowlist and blocklist: Let's discuss how the IP addresses are specified using the configuration. There are two parameters, namely blocklist and allowlist, that can be used to specify the location of files that contain IP addresses that need to be blocked and allowed, respectively. As noted earlier, the terms *blocklist* and *allowlist* are used interchangeably with *blacklist* and *whitelist*, respectively.

  An example IP blocklist file named ipbl.txt is shown in the following command. The file specifies one IP address per line. Correspondingly, the blocklist parameter will be configured to the name of this file. That is, the configuration line will be as follows:

  ```
  reputation = {
          blocklist = 'ipbl.txt'
  }
  ```

  The ipbl.txt file will be at that path specified in BLACK_LIST_PATH, and the contents of the ipbl.txt file will be IP addresses or CIDR entries, as shown next (note that BLACK_LIST_PATH can be specified as an absolute path or a relative path):

  ```
  192.168.1.1
  145.254.160.237
  ```

- priority: If an IP address has a match in both blocklist and allowlist, the higher priority list's decision will take precedence. This priority is specified using the priority parameter, which can take the values allowlist or blocklist.

  For example, consider the case when an entry '192.168.2.0/24' is specified in the blocklist file, while a specific IP address, '192.168.2.26', is specified in the allowlist file. Now, when a packet with IP address is being analyzed by the IP reputation inspector, there is a conflict. In such cases, the priority parameter determines the decision – to allow or block the packet.

- `nested_ip`: In the presence of network tunneling or **Generic Routing Encapsulation (GRE)** protocol packets, such as the IP-in-IP protocol, there are layers of IP headers. In such a case, the IP reputation inspector will need to know which IP address is the one that should be matched upon – inner, outer, or all. This is specified using the configuration parameter called `nested_ip`, which takes the values `outer`, `inner`, or `all`.

- `list_dir`: Using the `blocklist` and `allowlist` parameters, we can only specify one file for blocking and allowing. This works irrespective of which interface the packet was received on. The IP reputation inspector also supports specifying interfaces with blocklists and allowlists. This is specified using a special file called the *manifest* file. The name of the manifest file is hard-coded, `interface.info`, and it needs to be present in a directory that is configurable by the parameter called `list_dir`. Each line in the manifest will be in the following format:

  ```
  file_name, list_id, action (block, allow, monitor), [interface
  information]
  ```

  The interface information is optional, and if it is empty, the list will be applied to all interfaces.

  Let's take an example. In this case, `list_dir` is set as follows:

  ```
  reputation.list_dir = '../lists'
  ```

  In the `lists` directory, we would have a file named `interfaces.info`, which is the manifest file. A simple example of the manifest file is as follows:

  ```
  ipbl.txt, 1, block
  ```

  In this case, the file name is `ipbl.txt` and all the IP addresses listed in that file will be blocked. Since no interfaces are specified, this will be applied to all interfaces.

- `memcap`: The `memcap` configuration parameter is used to limit the memory used by the IP reputation inspector to store the IP addresses and relevant information. The value is an integer and will be specified in MB.

In the next section, let's discuss the functionality of the IP reputation inspector.

Functionality of the IP reputation inspector

The IP reputation module is implemented as an inspector module in Snort 3. Specifically, it is implemented as a network inspector. The functionality of this inspector is straightforward; we will discuss it in this section. Snort parses all the information and stores it in its memory using some efficient data structures (this is the IP reputation data). When the traffic is inspected, the inspector matches the source and destination IP addresses of the packet against the IP reputation data (that is, the blocklists and allowlists), and if there is a match, the corresponding decision is enforced (that is, the packet is dropped, allowed, or monitored).

Let's look at these processing stages in a bit more detail.

Data structure for storing IP reputation scores

The IP reputation module needs to store the IP addresses provided as blocklists and allowlists (for interfaces) in the memory, so that every analyzed packet can be compared against the stored IP addresses. During Snort's start up, the IP addresses are read and inserted into the data structure maintained by the IP reputation module. Thereafter, for every analyzed packet, the packet's IP addresses (source and destination) are compared (lookup operation). The insertion operation is done only when Snort starts up or at reconfigure time. The lookup operation is done for every packet that is processed. Therefore, for a high-speed network, the lookup operation has to be highly efficient in order to prevent performance issues.

Network routers also do similar lookups based on IP address; these are routing table lookups that are performed to route the packet to the correct destination.

Snort has implemented a routing table lookup for storing IP reputation data. This approach uses a multibit-trie method. The method performs longest prefix matching based on the IP address of the packet to determine whether there is a match in the stored IP reputation data.

The population of the IP reputation data is done during the configuration stage. The IP reputation configuration contains the following parameters, which specify filenames for IP blocklists and allowlists – `reputation.blocklist` and `reputation.allowlist`. In addition, the manifest file (`interface.info`) may contain more data to be processed.

Each of the blocklist or allowlist files contains IP addresses or CIDRs that are specified one per line. The processing of these files happens as follows:

1. Read one IP address or CIDR at a time.
2. A lookup operation is done to check whether an IP address or CIDR that matches the entry that is being added is already in the table. If yes, then an `IP_INSERT_DUPLICATE` error is presented.
3. The IP address or CIDR is inserted into the data structure using the insert API.

Once the Snort packet processing starts, the following steps will be done for each packet by the IP reputation inspector:

1. Get the interface info from the packet header.
2. Get the source IP address (and destination IP address) from the packet. For source and destination IP addresses, do:

   I.   Do a lookup against the IP reputation data that is stored in the trie.
   II. If a match is found, a decision is made whether to block or allow the packet, and an alert is generated.

Based on the `nested_ip` configuration setting, in the case of an IP-in-IP packet, the outer or inner IP address is retrieved. In case `all` is specified the lookup will be done iteratively across all the IPs.

Now that we have looked at the functionality of the IP reputation inspector, let's look at the various logs and alerts generated by the module.

IP reputation inspector – alerts and pegs

The IP reputation inspector can generate the following alerts, as shown by the `--help-module` command:

```
./build/src/snort --help-module reputation
Rules:
136:1 (reputation) packets blocked based on source
136:2 (reputation) packets trusted based on source
136:3 (reputation) packets monitored based on source
136:4 (reputation) packets blocked based on destination
136:5 (reputation) packets trusted based on destination
136:6 (reputation) packets monitored based on destination
```

We have to put a stub rule in the rules file to enable any of the preceding alerts that we need. For example, in order to enable alerting when the inspector blocks a packet based on source IP address, we will have this rule in our rules file:

```
alert ( gid:136; sid:1; msg:"(reputation) packets blocked based on
source"; priority:3; )
```

The IP reputation inspector module creates alerts when a packet is blocked. The alert will be created in the format specified in the Snort configuration. An alert for a packet that is blocked is as follows:

```
[**] [136:1:1] "(reputation) packets blocked based on source" [**]
[Priority: 3]
05/13-05:17:10.295515 145.254.160.237:3371 -> 216.239.59.99:80
TCP TTL:128 TOS:0x0 ID:3917 IpLen:20 DgmLen:761 DF
***AP*** Seq: 0x36C21E28  Ack: 0x2E6B5384  Win: 0x2238  TcpLen: 20
```

This was an alert that was created when the source IP matched an entry in the blocklist. Similarly, if the match is on the destination IP, then the message will be as follows:

```
[**] [136:4:1] "(reputation) packets blocked based on destination"
[**]
```

In addition to alerts, the IP reputation inspector also maintains various stats and counts called **pegs**. Here are the various pegs and what they track:

```
reputation.packets: total packets processed (sum)
reputation.blocked: number of packets blocked (sum)
reputation.trusted: number of packets trusted (sum)
reputation.monitored: number of packets monitored (sum)
reputation.memory_allocated: total memory allocated (sum)
reputation.aux_ip_blocked: number of auxiliary ip packets blocked
(sum)
reputation.aux_ip_trusted: number of auxiliary ip packets trusted
(sum)
reputation.aux_ip_monitored: number of auxiliary ip packets monitored
(sum)
```

The statistics on the pegs are listed as part of the Snort output during the process termination. The section of the output that is related to IP reputation as printed out during the Snort process termination is as follows:

```
reputation
                    packets: 26
                    blocked: 5
--------------------------------------------------------
Summary Statistics
--------------------------------------------------------
timing
                    runtime: 00:00:00
                    seconds: 0.035458
                  pkts/sec: 16639
                Mbits/sec: 20
o")~    Snort exiting
```

We have looked at the various alerts and logs that were generated by the IP reputation inspector in this section.

Summary

In this chapter, we discussed and learned about the IP reputation inspector module in Snort 3. We discussed how IP blocking evolved in Snort to the current state. We discussed how the IP reputation inspector module is configured, the various configuration parameters, and their importance. Finally, we discussed the functionality of the module and how it works.

The IP reputation inspector is a key module that can be extremely effective operationally. The effectiveness is as good as the quality of the IP blacklist. Although there are challenges in creating and maintaining a comprehensive and effective blacklist, there are strategies, such as inter-organization collaboration, that can help.

In the next chapter, we will look at Snort rules. Rules form the core of the Snort intrusion detection and prevention system. We will look at the syntax of Snort rules, and the various features that are available for a rule author.

Part 4:
Rules and Alerting

In the final section of the book, we will discuss Snort rules, including the structure, the syntax, and the details of rule headers and rule options. The chapter on Snort rules will familiarize you with Snort rules, enabling you to understand existing snort rules as well as create new ones. This final part of the book also discusses the alerting and logging capability of Snort 3. The various alert output plugins and their configuration are discussed in the chapter on *Alert Subsystem*. We will discuss the OpenAppID feature of Snort in *Chapter 15*. This module enhances the application detection capability of Snort, and it also extends it as a separate package so that it can be updated frequently as needed. The final chapter discusses various miscellaneous topics on Snort 3, including troubleshooting and the migration of a Snort 2 configuration to Snort 3.

This part has the following chapters:

- *Chapter 13, Rules*
- *Chapter 14, Alert Subsystem*
- *Chapter 15, OpenAppID*
- *Chapter 16, Miscellaneous Topics on Snort 3*

13
Rules

Snort is, at its core, a rules-based network intrusion detection/prevention system. Snort is a highly complex system and has several modules, such as Codecs and Inspectors that analyze the various protocols that traverse the network. All the analysis and processing that is done by Codecs and Inspectors are geared towards rule-based matching.

Snort rules are written to specify special network conditions or traffic patterns in order to detect and prevent attacks. Snort rules are written using a custom Snort syntax. The Snort Rules Engine parses the rules and matches the network traffic against the rules. This chapter will provide you with knowledge and details about the structure and syntax of a Snort rule, and about the different types of Snort rules. You will also learn about some Snort rule writing recommendations. In short, you will be able to write your own Snort rules.

The following topics will be covered:

- Snort rule – the structure
- Rule header
- Rule options
- Recommendations for writing good rules

Snort rule – the structure

An example Snort rule is as follows:

```
alert tcp $EXTERNAL_NET any -> $HOME_NET 80 (msg:"HTTP GET Request
- Inbound"; flow:established,to_server; http_method; content:"GET";
priority:5; sid:12345678;)
```

Basically, a Snort rule can be divided into two parts, namely the **rule header** and the **rule options**. The first part of the Snort rule (until the first parenthesis) is called the rule header. The part of the rule from the first parenthesis to the parenthesis at the end is called the rule options.

In the example rule, the rule header is as follows:

```
alert tcp $EXTERNAL_NET any -> $HOME_NET 80
```

The rule options are as follows:

```
(msg:"HTTP GET Request - Inbound"; flow:established,to_server; http_
method; content:"GET"; priority:5; sid:12345678;)
```

Before we delve into the rule header in detail, we will introduce the three types of rules introduced in Snort 3:

- Service rules
- File rules
- File identification rules

The main advantage of these new rule types is that they simplify the rule header.

Service rule

A service rule is a type of rule that has a rule header that clearly specifies what service (protocol) the rule applies to. For example, for a typical HTTP rule, the new service rule would look like this:

```
alert http (msg:"HTTP GET Request - Inbound"; flow:established,to_
server; http_method; content:"GET"; priority:5; sid:12345678;)
```

For this rule to be applied, the relevant session must be first identified as HTTP. This identification is done by the wizard inspector, and by the application identification inspector.

In this format, however, since the source and destination IP addresses are not specified, the rule will be applied in both directions – *inbound* and *outbound*. This may be important for organizations that may give different priority or importance to an alert that is detected in an inbound direction versus an outbound direction. It should be noted that the protocol can be specified in a traditional rule header as well.

File rule

A file rule is a rule that specifies an action and the `file` keyword in the rule header. The file rule will be applied to all files irrespective of the protocol that is used to transfer the file. Consider a PDF file that could be sent via email (using SMTP, POP, or IMAP protocol) or downloaded using browser (using HTTP protocol) or transferred using FTP. Irrespective of what mechanism is used, the same file rule will be applied against any file that is a PDF that arrives using the supported list of protocols (HTTP, SMTP, IMAP, POP3, SMB, and FTP).

Consider a case where we are interested in detecting the `malicious` pattern in a file that could be transferred by HTTP download or via SMTP. Without the file rule, we must have two rules. Here's the first:

```
alert tcp $EXTERNAL_NET $HTTP_PORTS -> $HOME_NET any (msg:"Malicious
File detected over HTTP - Inbound"; flow:established,to_client; file_
data; content:"malicious"; service:http; priority:5; sid:12345777;)
```

This is the second rule:

```
alert tcp $EXTERNAL_NET any -> $HOME_NET 25 (msg:"Malicious File
detected via SMTP - Inbound"; flow:established,to_server; file_data;
content:"malicious"; service:smtp; priority:5; sid:12345778;)
```

These can be replaced by a single file rule, as follows:

```
alert file (msg:"Malicious File detected - Inbound"; file_data;
content:"malicious"; priority:5; sid:12345678;)
```

It should be noted that the file rule should not specify the `flow` keyword in the rule options since the concept of flow does not apply to the context of a file. Similarly, the `service` keyword should not be used since the file has to be detected irrespective of the protocol.

File identification rule

A file identification rule is a rule that specifies an action and the `file_id` keyword in the rule header. This is a special rule – while other rules are aimed at detecting the presence or absence of malicious behavior in the network traffic, this rule is aimed at identifying a file.

In order to enable the feature, the following configuration needs to be included in the Snort configuration file. The `file_id` parameter specifies the location of the magic file:

```
file_id = { rules_file = 'file_magic.rules' }
file_policy = { }
```

The `file_magic` file includes rules such as the following that identify particular files. Each of these rules should use a unique `id` (the `id` value is `100` for this rule):

```
file_id (msg:"PDF File Detected"; file_data; content:"%PDF-"; depth:5;
offset:0; file_meta:type PDF, id 100, category "PDF file"; gid:4;
sid:123;)
```

The result of this identification is used for further analysis. In this rule, when the rule triggers for a particular session, Snort also marks the file as a PDF (using the `file_meta` rule option). There is another rule option called `file_type`. The `file_type` rule option checks if the identified file is a particular type:

```
alert file (msg:" PDF File Inbound"; file_type: PDF;
content:"|2F|Type|20|EmbeddedFile"; sid:234; )
```

Identifying the file type and then applying relevant rules against the file content works well and reduces false positives.

Next, we will look at the rule header portion of a Snort rule. We will discuss the syntax of the rule header in detail.

Rule header

We saw in the previous section the basic structure of a Snort rule, and the two top-level parts of a rule, namely rule header and rule options. Based on the rule header structure, rules are classified as traditional, service, file, and file_id. We looked at service, file, and file_id rules in the previous section. In this section, let's look at the traditional rule header structure.

Traditional rule header

For traditional rules, the rule header structure is as follows:

```
action protocol source_ip source_port directionality destination_ip
destination_port
```

Action

The `action` or rule action specifies to Snort what to do when there is a successful match for the rule. The `action` verb can take any of the following values:

- `alert`: The `alert` action causes an alert to be generated and also the corresponding packet to be logged.

- `drop`: The `drop` action generates an alert, and the corresponding packet to be dropped (when Snort is running in inline IPS mode).

- `block`: The `block` action works exactly like `drop`, but in addition to dropping the packet, it also blocks all the subsequent packets in the flow.

- `log`: The `log` action causes the current packet to be logged.

- `pass`: The `pass` action causes the current packet to be passed.

These actions do not actively respond to the network traffic that triggered a rule. The next few actions may be considered as some sort of response:

- `react`: The `react` action results in sending a response to the client side, followed by the termination of the session.
- `reject`: The `reject` action causes Snort to terminate the session by sending TCP RST segment or ICMP unreachable datagram.
- `rewrite`: The `rewrite` action causes the data in the current packet to be rewritten.

Snort 3 has the following modes of operation – inline, tap, and inline-test. This is configured via the configuration option called `config policy_mode`. This setting affects how the action is interpreted. For example, a `drop` rule in tap mode would be like an `alert` rule.

Protocol

The `protocol` specifies the network or transport layer protocol, or the identified service.

The network or transport protocols supported are as follows:

- `ip`
- `tcp`
- `udp`
- `icmp`

In addition, the following services may be specified as well:

- `http`
- `smtp`
- `imap`
- `pop`

The `protocol` field is used to categorize the rules to match the incoming traffic appropriately.

Source IP

The `source_ip` field represents the source IP address of the network traffic. The source IP address of the traffic must match the source IP value specified in the rule for the rule to match.

This field can be a single IP address, a subnet, or a list of IP addresses and subnets. It is called an **IP list**. An example IP list is `[192.168.1.0/24,172.16.1.1]`.

In addition, the IP list can have negated IP addresses and/or subnets. Both IPv4 and IPv6 addresses are supported by Snort 3.

It is common to create variables for various IP lists. For example, HOME_NET is a common variable name that is used for an IP list containing all the individual IP addresses and IP ranges that constitute a home network.

Source port

The source_port field represents the TCP or UDP source port. The value can be a list of individual ports, port ranges, and negated ports. An example port list is [80, 3128, 8000:8100, !8050]. This list includes the individual ports 80 and 3128, and the ports between 8000 to 8100 except 8050.

Directionality

The directionality operator in a rule shows the direction of the traffic that matches the rule. The directionality values are -> and <>. The -> value shows that the rule matches a particular direction of the traffic, whereas <> shows that the rule can match either direction.

Destination IP

The destination_ip field represents the destination IP address of the network traffic. The destination IP address of the traffic must match the destination IP value specified in the rule for the rule to match.

This field can be a single IP address, a subnet, or a list of IP addresses and subnets. It is called an IP list. An example IP list is [192.168.1.0/24,172.16.1.1].

In addition, the IP list can have negated IP addresses and/or subnets. Both IPv4 and IPv6 addresses are supported by Snort 3.

Destination port

The destination_port field represents the TCP or UDP destination port. The value can be a list of individual ports, port ranges, and negated ports. An example port list is [80, 3128, 8000:8100, !8050]. This list includes the individual ports 80 and 3128, and the ports between 8000 to 8100 except 8050.

It is common to create variables for various port lists. For example, HTTP_PORTS is a common port variable that contains the various ports that have HTTP protocol associated.

That covers the Snort rule header part, and now we can move on to discuss the Snort rule options.

Rule options

The rule options constitute the second half of a Snort rule. The rule header is mostly about the network and transport header of the network packet. On the other hand, the rule options generally deal with the characteristics of the payload (the contents of the payload, the size of the payload), the state of the protocol or the session, and so on. However, there is also a set of rule options that deal with rule metadata, such as rule message, signature ID, revision, and priority. The Snort rule options are evaluated from left to right. In this section, we will look at the various rule options.

General rule options

The general rule options within Snort serve the purpose of providing supplementary information about the rule. These options do not directly impact the detection functionality, but rather are designed to enhance the understanding and management of rules within the Snort system. Let's take a closer look at each of these options:

- `msg`: The `msg` option allows the rule author to include a descriptive message that provides additional context or details about the rule. This message helps to document the purpose or intent of the rule, making it easier for administrators and analysts to comprehend its meaning and significance.

- `reference`: The `reference` option enables the rule author to include external references, such as URLs or IDs, that provide more information about the threat or vulnerability addressed by the rule. These references can serve as valuable resources for analysis, investigation, and further understanding of the associated threat.

- `sid`: The `sid` (signature ID) option assigns a unique numerical identifier to each rule. This identifier serves as a reference point for identifying and managing specific rules within the Snort system. The `sid` value helps administrators and analysts with rule tracking, correlation, and reporting, allowing efficient rule management and analysis.

- `rev`: The `rev` (revision) option indicates the version or revision number of a particular rule. It enables tracking changes and updates to rules over time, aiding in rule management, documentation, and troubleshooting. By keeping track of rule revisions, administrators can easily identify and manage rule modifications or updates.

- `classtype`: The `classtype` option allows the rule author to assign a classification type or category to the rule. This classification helps organize and prioritize rules based on the type of threat or attack they are designed to detect. It assists with filtering and focusing on specific rule categories during analysis and response, improving the efficiency of the detection and response process.

- priority: The priority option assigns a numerical value to indicate the importance or severity of a rule. It helps with prioritizing alerts and determining the order in which they should be addressed. By assigning higher priority values to more critical rules, administrators can ensure that immediate attention is given to the most severe threats or attacks.

These general rule options play a crucial role in providing additional information, organization, and management capabilities within the Snort intrusion detection system. By utilizing these options effectively, administrators and analysts can gain better insights into the rules, facilitate rule tracking, and prioritize the response to detected threats.

Payload options

The payload detection rule options provide a powerful means of specifying conditions for inspecting and identifying malicious or suspicious content within network packet payloads. The various rule options enable the rule author to define specific criteria to match against various aspects of packet payloads. These rule options support advanced techniques such as regular expressions and byte-level matching, enabling the detection of complex attack patterns. These rule options can help with the detection and prevention of various types of cyber threats, such as malware, exploits, and data exfiltration attempts, enhancing the overall security posture of the network.

The payload rule options include the following.

content

The content rule option is the most widely used payload-based rule option. It is a simple yet very powerful and versatile rule option. The content rule option takes arguments, so it must be followed by a : and then by the *pattern*.

For example, if we would like to search for the foobar pattern in the packet payload, we would use the content rule option as follows:

```
content:"foobar";
```

In addition to alphanumerical characters, the content option can search for a binary pattern or a combination of alphanumeric and binary characters. Binary patterns have to be enclosed within | characters:

```
content:"HTTP|2F|1|2E|1";
```

This rule option would search for the pattern HTTP/1.1.

By default, the content rule option searches for the pattern in the entire payload or buffer that is under analysis. However, there are content modifiers that can be combined with the rule option value that will alter the limits of the search. These content modifiers are offset, depth, distance, and within.

There is another content modifier called `nocase`, which alters the behavior of the search. This will search for the pattern by ignoring whether the letters are upper or lower case.

offset

This modifier allows you to specify the starting position within the payload where Snort should begin searching for the content. It is helpful when the content of interest is located at a specific position within the payload.

depth

This modifier specifies the *depth* for the search. In other words, it specifies the number of bytes that Snort will search for the specified content. This can be useful when you are only interested in a specific portion of the payload or want to optimize performance by reducing the search scope.

distance

This modifier is used in conjunction with the `content` modifier to specify a distance from a previous match where Snort should start searching for subsequent matches. This is useful for finding multiple occurrences of the content with a specific separation.

within

This modifier is used in conjunction with the `content` modifier to specify a limit to the search from a previous match until where Snort should search for subsequent matches.

Before we look at some examples of rules using the `content` rule option and the `content` modifiers, let's learn about Snort's inspection buffers and the concept of sticky buffers.

Inspection buffers

As the network traffic is analyzed by the relevant inspectors, Snort populates various inspection buffers so that the rule matching can be done within the right context, resulting in better efficiency and better accuracy.

For example, as the HTTP request is processed by the HTTP inspector, various inspection buffers that are related to the HTTP request are populated. For example, the HTTP request is extracted into the HTTP method buffer, the HTTP request URI populates the HTTP URI buffer, and the HTTP header populates the HTTP header buffer.

The following is an extensive list of HTTP-related inspection buffers:

- `http_uri`: This buffer will contain the normalized HTTP URI.
- `http_raw_uri`: This buffer will contain the unnormalized HTTP URI.
- `http_header`: This buffer will contain the normalized HTTP headers.

- `http_raw_header`: This buffer will contain the unnormalized HTTP headers.

- `http_cookie`: This buffer will contain the normalized HTTP cookies.

- `http_raw_cookie`: This buffer will contain the unnormalized HTTP cookies.

- `http_client_body`: This buffer will contain the normalized HTTP request body.

- `http_raw_body`: This buffer will contain the unnormalized HTTP request body and response data.

- `http_param`: This buffer will contain the specific HTTP parameter values.

- `http_method`: This buffer will contain the HTTP request methods.

- `http_version`: This buffer will contain the HTTP request and response versions.

- `http_stat_code`: This buffer will contain the HTTP response status codes.

- `http_stat_msg`: This buffer will contain the HTTP response status messages.

- `http_raw_request`: This buffer will contain the unnormalized HTTP start lines.

- `http_raw_status`: This buffer will contain the unnormalized HTTP status lines.

- `http_trailer`: This buffer will contain the normalized HTTP trailers.

- `http_raw_trailer`: This buffer will contain the unnormalized HTTP trailers.

- `http_true_ip`: This buffer will contain the original client IP address as stored in various request proxy headers. The header options include `X-Forwarded-For` and `True-Client-IP`.

- `http_version_match`: This is the non-sticky buffer option used to match the HTTP message's version by comparing against a list of versions.

- `http_num_headers`: This is the non-sticky buffer option used to test the number of HTTP headers. The rule option enables a comparison of the number of HTTP headers against a specific value or a range of values.

- `http_num_trailers`: This is the non-sticky buffer option used to test the number of HTTP trailers. The rule option enables a comparison of the number of HTTP trailers against a specific value or a range of values.

- `http_num_cookies`: This is the non-sticky buffer option used to test the number of HTTP cookies against a specific value or a range of values

Next, let us look at the sticky buffer feature in Snort.

Sticky buffers

By default, rule options such as `content` are evaluated against the packet payload (which is stored in the `pkt_data` buffer). All subsequent content matches will also be performed against the same buffer. In order to specify to Snort that the content matching has to be done against a different buffer, for example, the HTTP URI buffer, we have to specify that in the rule. This is specified by stating `http_uri;` before the `content` rule option. The concept of a sticky buffer is that once the analysis buffer is changed by specifying a buffer name, all the subsequent analysis will be done on that buffer. In other words, all analysis will stick to the specified buffer.

The following are rule options that are related to various sticky buffers:

- `pkt_data`: The `pkt_data` rule option changes the sticky buffer to the default packet payload (the normalized packet data buffer).

- `raw_data`: The `raw_data` rule option changes the sticky buffer to the unnormalized packet payload.

The use of sticky buffers enables more accurate detection and also makes the matching process more efficient.

fast_pattern

As mentioned earlier, the Snort rule options are evaluated from left to right. However, it should be noted that Snort does not evaluate all the HTTP rules completely (in a left-to-right fashion). In fact, Snort only evaluates a subset of the rules for which the multi-pattern search returns a successful match. A special pattern from each rule is selected for a multi-pattern search phase. This pattern for each rule is specified using the `fast_pattern` rule option. (For rules that do not specify a special pattern using the `fast_pattern` option, the longest `content` pattern will be used as a special pattern for this.) These patterns are stored in Snort so that any buffer can be matched against all the patterns with a single search, hence the name multi-pattern search. String algorithms such as *Aho-Corasick string matching* are used for this purpose. In the multi-pattern search phase, only a subset of the patterns will return a successful match. Only the rules corresponding to these special patterns are evaluated completely.

Next, let's look at the `pcre` rule option, which is a very powerful rule option for detecting patterns.

pcre

The name **pcre** is an abbreviation for **Perl Compatible Regular Expression**. The `pcre` rule option matches the specified regular expression against the payload or buffer data. As the name suggests, the regular expression specified using the `pcre` rule option must be compatible with Perl regular expression syntax.

The rule option syntax is as follows:

```
pcre:"/regular expression here/";
```

The concept of sticky buffer applies to the pcre option as well. For example, the following combination of rule options uses the http_uri sticky buffer option to set the DOE pointer to the HTTP URI buffer before applying the pcre option.

```
http_uri; pcre:"/\x2Ephp\x3Eparam\x3D/";
```

The pcre rule option also supports a set of flags that alter the regular expression evaluation. The flags are specified by lowercase or uppercase letters listed after the ending forward slash in the pcre rule option:

```
pcre:"/regular expression here/flags";
```

The following are the flags that are applicable for the pcre rule option:

- i: The i flag makes the regex case insensitive, allowing it to match uppercase and lowercase letters without distinction.

- s: When the s flag is enabled, the dot metacharacter in the regex pattern also matches newline characters.

- m: The m flag alters the behavior of ^ and $ anchors to match the beginning and end of lines, in addition to the start and end of the entire string.

- x: By using the x flag, whitespace characters in the regex pattern are ignored unless escaped or inside a character class, making the pattern more readable.

- A: The A flag ensures that the regex pattern matches only at the start of the buffer, like using the ^ character anchor.

- E: When the E flag is set, the $ anchor in the regex pattern matches only at the end of the subject string, instead of matching the end of lines or the entire string.

- G: The G flag reverses the default greediness of quantifiers in the regex pattern. They become non-greedy by default but can become greedy if followed by ?.

- O: The O flag overrides the default PCRE match limit and match limit recursion settings for the specific expression, allowing you to modify them as needed.

- R: By using the R flag, the regex search starts from the end of the last match instead of the beginning of the buffer, facilitating multiple matches on the same string.

The pcre option integrates the power of regular expressions to the rule-matching process. This is a very powerful and useful option; however, it is also computationally expensive. Therefore, rule authors should use their judgment when using this option.

dsize

The `dsize` rule option is used to check the size of the packet payload. The rule option supports the >, <, >=, <=, and <> operators. The absence of any operator indicates the = sign. Here's some examples:

- `dsize:20;`: Checks for a packet with payload size equal to 20.
- `dsize:>20;`: Checks for a packet with payload size greater than 20.
- `dsize:>=20;`: Checks for a packet with payload size greater than or equal to 20.
- `dsize:10<>20;`: Checks for a packet with payload size between 10 and 20.

The `dsize` option is used to check the size of the payload for regular datagrams. It does not work with reassembled stream payloads. The `stream_size` option should be used in such cases. The syntax for `stream_size` is exactly the same as for `dsize`; the only difference is that it applies to reassembled streams.

bufferlen

The `bufferlen` rule option is used to check the size of the relevant buffer. By default, the buffer is `pkt_data` (the packet payload data buffer), but by using sticky buffers we can check the size of any buffer. For example, the following combination checks for the size of the HTTP URI buffer:

```
http_uri; bufferlen:<100;
```

Similar to `dsize`, the =, <, >, >=, <=, and <> operators are supported.

isdataat

The `isdataat` rule option is used to check whether there is data at a particular byte count. When used in an absolute sense, the option checks for the presence of data from the start of the buffer. When used in a relative sense, the option checks for the presence of data from the point of reference:

```
isdataat:[!]location_index[,relative]
```

For example, `isdataat:100` would check for the presence of data at an offset of 100 bytes from the start of the `pkt_data` buffer.

Non-payload options

Non-payload rule options in Snort provide the ability to analyze and match specific attributes of network packets that are not related to the payload's content. These options enhance the flexibility and effectiveness of rule creation, allowing a comprehensive examination of network traffic. Let's look at some examples of non-payload rule options in Snort.

IP header options

These options focus on conditions based on IP header fields, such as IP protocol type, **Time-to-Live** (**TTL**), and IP fragmentation flags. Let's look at the rule options that come under this category.

ttl

The `ttl` rule option is used to check the IP TTL value. For example, we could match the subset of packets that have a TTL value of less than 10 as follows:

```
ttl:<10;
```

The syntax for the rule option is as follows:

```
ttl:[<|>|=|!|<=|>=]value;
ttl:min_val{<>|<=>}max_val};
```

The `ttl` field in the IP header uses 8 bits. Therefore, the maximum value of the TTL is 255.

ip_proto

The `ip_proto` rule option is used to match against the IP protocol value specified in the IP header. Some example IP protocols and their protocol numbers are as follows:

- `1` – **Internet Control and Messaging Protocol (ICMP)**.
- `2` – **Internet Group Management Protocol (IGMP)**.
- `6` – **Transport Control Protocol (TCP)**.
- `11` – **User Datagram Protocol (UDP)**.
- `50` – **Encapsulating Security Protocol (ESP)**.
- `51` – **Authentication Header (AH)**.

So, the `ip_proto: 50;` rule option would match traffic that uses ESP protocol (which is part of the IPSec protocol suite).

fragbits

The `fragbits` rule option is used to match against IP fragmentation flags. There are 3 bits in the IP header's third byte, which are used as follows:

- `M`: This bit indicates whether there are more fragments.
- `D`: This bit specifies that the datagram should not be fragmented.
- `R`: This is a reserved bit.

The `fragbits` rule option can be used to match these bits. For example, `fragbits:M;` will match packets that have the *More Fragments* bit.

fragoffset

This rule option checks for the IP fragmentation offset field value. The rule option syntax supports the operators such as =, >, <, >=, <=, and <>, which are used to compare numbers. This rule option checks whether the IP fragmentation value for the packet is 0. This is useful for detecting the first IP fragment:

```
fragoffset:0;
```

This rule option checks whether the IP fragmentation value for the packet is not 0. This is useful for detecting the IP fragments with a non-zero offset (all the other fragments than the first fragment):

```
fragoffset:!0;
```

ICMP-related options

These options allow you to match specific **Internet Control Message Protocol** (**ICMP**) message types or codes, enabling the detection of ICMP-based attacks or abnormal ICMP behavior.

The rule options – `itype`, `icode`, `icmp_id`, and `icmp_seq` – are concerned with ICMP header field values. The rule option syntax supports the operators such as =, >, <, >=, <=, and <>, which are used to compare numbers.

TCP/UDP header options

Rule options related to TCP/UDP headers include source and destination port numbers, TCP flags (for example, `SYN`, `ACK`, and `FIN`), and UDP packet length. They are useful for detecting specific network services or identifying TCP/UDP-based attacks.

Flow options

Snort supports rule options for flow-based analysis, including checking for established connections, tracking packet direction (for example, `to_server` or `to_client`), and examining flow direction based on port numbers. Flow options help identify and analyze network flows.

Metadata options

Snort allows the inclusion of metadata in rules, providing additional contextual information about the rule or the network traffic being analyzed. Metadata options enhance rule management and analysis by adding descriptive information or organizational tags.

Next, let's see some good rule-writing recommendations. These are good tips to keep in mind when creating typical Snort rules.

Recommendations for writing good rules

In this section, we will look at a few of the tips or recommendations for writing good Snort rules.

Using fast_pattern wisely

The use of `fast_pattern` is a key factor that affects the Snort's runtime performance. The right or wrong choice of `fast_pattern` can have an impact on the performance of Snort.

Choosing the most unique pattern for `fast_pattern` is very crucial. If the pattern that is chosen as the `fast_pattern` is not unique but rather very common, then the pattern will always match all traffic and will result in unnecessary evaluation of the rule.

Using the inspection buffers for rule matching

When the pattern that we are looking for is available in an inspection buffer, always search in the buffer rather than the entire packet payload. Since the buffer will contain a subset of the entire packet payload, the efficiency of the search will be improved.

Defining the right service or protocol

Always write rules for a specific service or protocol when it is possible. For example, the HTTP protocol works on top of the TCP protocol. When we can write a rule as an HTTP rule, we should choose that option instead of writing it as a TCP rule.

Summary

Snort rules form the backbone of effective network intrusion detection and prevention. In this chapter, we explored the key components and concepts of Snort rules, covering both payload and non-payload options. We delved into the various rule options available in Snort, such as content matching, modifiers, and non-payload attributes. We also looked at some of the tips and pointers to keep in mind when writing Snort rules.

Understanding the intricacies of Snort rules will enable you to design precise and tailored rulesets that efficiently identify and mitigate network threats. The next chapter will deal with Snort's alert subsystem.

14
Alert Subsystem

The alert subsystem is one of the key components of Snort. The goal of the Snort system is to inspect the network traffic and identify (and stop) malicious traffic. To do that, the traffic is first captured (by DAQ modules), then decoded (by decoder modules), analyzed (by inspector modules), and matched against the signatures (by rules module). In this chapter, we will discuss what happens when there is a successful match for a signature. We will discuss the role of the alert subsystem, that is, creating an alert when there is a successful identification of a malicious packet or session.

At a high level, we will study the Snort alert subsystem, how it works, the various types or formats of alerts, and the configuration parameters.

We will be covering the following main topics:

- Post-inspection processing
- Alert formats

Post-inspection processing

In this section, we will discuss what happens after the packet inspection and rule matching is complete. When there is a successful rule match, an event is generated. The event is then checked against any applicable thresholding. Subsequently, the rule action is applied to the packet. Finally, an alert is also created for the event.

The following steps are involved when Snort successfully triggers a rule and generates an event:

- Event generation
- Check event thresholding rules
- Apply rule action for the packet
- Log the alert

Let's discuss each of these steps briefly.

Event generation

A Snort event is associated with a corresponding rule that was triggered; subsequently, that event is represented by a generator ID (`gid`) and signature ID (`sid`). A typical Snort rule has an associated signature ID, which is specified using the rule option called **sid** (please refer to *Chapter 13*, for more details). The `gid` for a typical rule is `1`. Events can be generated from any module, for example, inspectors or decoder modules. Each of these modules has a unique `gid`.

When an event is generated, it is placed internally in an Event Queue data structure. At the end of the packet processing, a subset of these events is logged as alerts. There is a configurable limit to how many events can be placed in the queue, and there is another configurable setting to how many of the queued events are logged as alerts. The maximum value for both these configurable settings is `4294967295` (the maximum value a 32-bit unsigned integer can hold, which is denoted as `max32`):

```
int event_queue.max_queue = 8: maximum events to queue { 1:max32 }
int event_queue.log = 3: maximum events to log { 1:max32 }
```

By default, a total of 8 events can be queued in the Event Queue per packet. Also, a total of 3 events will be logged as alerts from the queued events per packet.

The criteria to be used for ordering events in the Event Queue, when more than one event is generated per packet, is specified using the `order_events` configuration parameter. Currently, the two options are `priority` and `content_length`:

```
enum event_queue.order_events = 'content_length': criteria for
ordering incoming events { 'priority|content_length' }
```

Every Snort rule has an action, such as `drop`, `alert`, `log`, and `pass`. All the rules that have the `drop` action are in the same action group. Snort evaluates the rules of different action groups one after the other. `process_all_events` controls whether an event must be generated for all the action groups, or whether the processing can be stopped as soon as one of the action groups causes an event to be generated:

```
bool event_queue.process_all_events = false: process just the first
action group or all action groups
```

This is how events can be generated from the various modules of Snort, and based on the module that generates the event, the `gid` of the event changes. An event can be mapped to the module that generated the event using the `gid`. Now, let's discuss event thresholding.

Event thresholding

Event thresholding is a feature in Snort where certain types of events (specified by `sid`/`gid`) can be rate limited. This is very helpful for limiting the volume of alerts generated. Once an event is generated, it is checked against the event-filter-based rules before an alert is generated.

Here is a snippet of the `event_filter` configuration from the `snort.lua` configuration file that is shipped with the Snort package:

```
event_filter =
{
    -- reduce the number of events logged for some rules
    { gid = 1, sid = 1, type = 'limit', track = 'by_src', count = 2,
seconds = 10 },
..
}
```

This event filter limits the number of alerts generated from the rule with a `gid` of `1` and a `sid` of `1` to `2` per `10` seconds (when tracked by the source IP address).

Let's consider another example of a rule that we discussed in *Chapter 13*:

```
alert tcp $EXTERNAL_NET $HTTP_PORTS -> $HOME_NET any (msg:"Malicious
File detected over HTTP - Inbound"; flow:established,to_client; file_
data; content:"malicious"; service:http; priority:5; sid:12345777;)
```

If this rule is creating too many alerts, and if we need to limit the number of alerts that are created from this rule to 5 per minute per IP address, we could add the following `event_filter` to the configuration:

```
    { gid = 1, sid = 12345777, type = 'limit', track = 'by_src', count
= 5, seconds = 60 }
```

Event thresholding is practically very useful, especially when a signature is useful but at the same time noisy. Now, let's discuss the application of the action once a rule matches.

Applying a rule action to a packet

The next step is to apply a rule action to the packet, such as `drop` or `block`. This step is more useful for IPS mode, where there are several rule actions, such as `alert`, `drop`, `block`, `reject`, and `replace`. The relevant options in IDS mode are `alert`, `log`, and `pass`.

The following is a rule with the `alert` action:

```
alert tcp $EXTERNAL_NET $HTTP_PORTS -> $HOME_NET any (msg:"Malicious
File detected over HTTP - Inbound"; flow:established,to_client; file_
data; content:"malicious"; service:http; priority:5; sid:12345777;)
```

Here, we have a rule with the `drop` action:

```
drop tcp $EXTERNAL_NET $HTTP_PORTS -> $HOME_NET any (msg:"Malicious
File detected over HTTP - Inbound"; flow:established,to_client; file_
data; content:"malicious"; service:http; priority:5; sid:12345777;)
```

The difference between `drop` and `block` is that in the case of a `drop` rule, only the packet that triggers the alert is dropped, whereas in the case of a `block` rule, the packet that triggers the rule and all the subsequent packets belonging to that session are dropped.

On the other hand, the `reject` and `react` rule actions both involve Snort sending response packets to the host to terminate the session. In the case of a `reject` rule, if the session is a TCP session, Snort sends a TCP RST segment to terminate the session.

Now that we know how events are generated, let's discuss how an alert is generated from the relevant events.

Logging the alert

Finally, an alert is created for the event. An alert can be written in various formats. Here is a list of different alert formats:

- Unified2 format
- CSV format
- Alert Fast format
- Alert Full format
- JSON format
- Syslog format
- Talos format
- Unix sock format

We will discuss the most commonly used formats in the next section.

Alert formats

Let's discuss the different alert formats. To do that, let's take an example. We have a Snort signature, as follows:

```
alert http any any -> any any (msg:"Download HTML Rule";
flow:established,to_server; http_uri; content:"|2F|download|2E|html";
http_header; content:"Host|3A 20|www|2E|ethereal|2E|com";
content:"User|2D|Agent|3A 20|Mozilla",distance 0; sid:123459991;)
```

The signature is an HTTP signature that looks for `/download.html` in the HTTP URI, and `Host: www.ethereal.com` and `User-Agent: Mozilla` in the HTTP headers.

> **Note**
>
> Please note that the content rule option uses the | character to specify hexadecimal values. For example, |2D| would denote a - character.

The packet capture that we use for this exercise is given here (in tcpdump format). We note that all the criteria for the aforementioned Snort rule are satisfied by the HTTP request in the packet capture. Hence, we expect an alert from Snort:

```
06:17:08.222534 IP 145.254.160.237.3372 > 65.208.228.223.80: Flags [P.], seq 1:480,
ack 1, win 9660, length 479: HTTP: GET /download.html HTTP/1.1
        0x0000:  4500 0207 0f45 4000 8006 9010 91fe a0ed  E....E@.........
        0x0010:  41d0 e4df 0d2c 0050 38af fe14 114c 618c  A....,.P8....La.
        0x0020:  5018 25bc a958 0000 4745 5420 2f64 6f77  P.%..X..GET./dow
        0x0030:  6e6c 6f61 642e 6874 6d6c 2048 5454 502f  nload.html.HTTP/
        0x0040:  312e 310d 0a48 6f73 743a 2077 7777 2e65  1.1..Host:.www.e
        0x0050:  7468 6572 6561 6c2e 636f 6d0d 0a55 7365  thereal.com..Use
        0x0060:  722d 4167 656e 743a 204d 6f7a 696c 6c61  r-Agent:.Mozilla
        0x0070:  2f35 2e30 2028 5769 6e64 6f77 733b 2055  /5.0.(Windows;.U
        0x0080:  3b20 5769 6e64 6f77 7320 4e54 2035 2e31  ;.Windows.NT.5.1
        0x0090:  3b20 656e 2d55 533b 2072 763a 312e 3629  ;.en-US;.rv:1.6)
        0x00a0:  2047 6563 6b6f 2f32 3030 3430 3131 330d  .Gecko/20040113.
        0x00b0:  0a41 6363 6570 743a 2074 6578 742f 786d  .Accept:.text/xm
        0x00c0:  6c2c 6170 706c 6963 6174 696f 6e2f 786d  l,application/xm
        0x00d0:  6c2c 6170 706c 6963 6174 696f 6e2f 7868  l,application/xh
        0x00e0:  746d 6c2b 786d 6c2c 7465 7874 2f68 746d  tml+xml,text/htm
        0x00f0:  6c3b 713d 302e 392c 7465 7874 2f70 6c61  l;q=0.9,text/pla
        0x0100:  696e 3b71 3d30 2e38 2c69 6d61 6765 2f70  in;q=0.8,image/p
        0x0110:  6e67 2c69 6d61 6765 2f6a 7065 672c 696d  ng,image/jpeg,im
        0x0120:  6167 652f 6769 663b 713d 302e 322c 2a2f  age/gif;q=0.2,*/
        0x0130:  2a3b 713d 302e 310d 0a41 6363 6570 742d  *;q=0.1..Accept-
        0x0140:  4c61 6e67 7561 6765 3a20 656e 2d75 732c  Language:.en-us,
        0x0150:  656e 3b71 3d30 2e35 0d0a 4163 6365 7074  en;q=0.5..Accept
        0x0160:  2d45 6e63 6f64 696e 673a 2067 7a69 702c  -Encoding:.gzip,
        0x0170:  6465 666c 6174 650d 0a41 6363 6570 742d  deflate..Accept-
        0x0180:  4368 6172 7365 743a 2049 534f 2d38 3835  Charset:.ISO-885
        0x0190:  392d 312c 7574 662d 383b 713d 302e 372c  9-1,utf-8;q=0.7,
        0x01a0:  2a3b 713d 302e 370d 0a4b 6565 702d 416c  *;q=0.7..Keep-Al
        0x01b0:  6976 653a 2033 3030 0d0a 436f 6e6e 6563  ive:.300..Connec
        0x01c0:  7469 6f6e 3a20 6b65 6570 2d61 6c69 7665  tion:.keep-alive
        0x01d0:  0d0a 5265 6665 7265 723a 2068 7474 703a  ..Referer:.http:
        0x01e0:  2f2f 7777 772e 6574 6865 7265 616c 2e63  //www.ethereal.c
        0x01f0:  6f6d 2f64 6576 656c 6f70 6d65 6e74 2e68  om/development.h
0x0200:  746d 6c0d 0a0d 0a                               tml....
```

Figure 14.1 – The HTTP request used for triggering the example rule

The Snort command used is as follows:

```
snort3 -c snort.lua -l ~/Downloads/log/ -R ~/Downloads/Snort3/rules/
local.rules --daq dump -Q -r ~/Downloads/PCAPS/http.cap
```

The signature mentioned earlier in this section is defined in the rules file – `local.rules`. The `snort.lua` configuration file has a section where we specify what format we need for our alerts. Alternatively, Snort also supports a command-line option (`-A`) that may be used to specify the alert format.

Now, let's take a closer look at a few of the Snort alert formats, and we will start with CSV format.

CSV format

The module that creates alerts in CSV format is called the `alert_csv` module. The `alert_csv` module may be enabled via Snort command-line configuration or via configuration files.

- Command line: `-A alert_csv`
- Configuration file line: `alert_csv = {file = true, limit = 1000000}`

The `alert_csv` format has the following configuration parameters:

```
bool alert_csv.file = false: output to alert_csv.txt instead of stdout
multi alert_csv.fields = 'timestamp pkt_num proto pkt_gen pkt_len dir
src_ap dst_ap rule action': selected fields will be output in given
order left to right { action | class | b64_data | client_bytes |
client_pkts | dir | dst_addr | dst_ap | dst_port | eth_dst | eth_len |
eth_src | eth_type | flowstart_time | geneve_vni | gid | icmp_code |
icmp_id | icmp_seq | icmp_type | iface | ip_id | ip_len | msg | mpls |
pkt_gen | pkt_len | pkt_num | priority | proto | rev | rule | seconds
| server_bytes | server_pkts | service | sgt| sid | src_addr | src_ap
| src_port | target | tcp_ack | tcp_flags | tcp_len | tcp_seq | tcp_
win | timestamp | tos | ttl | udp_len | vlan }
int alert_csv.limit = 0: set maximum size in MB before rollover (0 is
unlimited) { 0:maxSZ }
string alert_csv.separator = ', ': separate fields with this character
sequence
```

As you can see, the `fields` parameter can be used to specify which fields need to be logged as part of the CSV alert. The default fields are as follows: `timestamp`, `pkt_num`, `proto`, `pkt_gen`, `pkt_len`, `dir`, `src_ap`, `dst_ap`, `rule`, and `action`.

Exercise

The line within the `snort.lua` configuration for alerts in CSV format is as follows:

```
alert_csv = {file = true, limit = 1000000}
```

Let's use a sample packet capture and run it through Snort so that it generates an alert in CSV format. The command line used in this exercise is as follows:

```
snort3 -c snort.lua -l ~/Downloads/log/ -R ~/Downloads/Snort3/rules/
local.rules --daq dump -Q -r ~/Downloads/PCAPS/http.cap
```

The CSV alert was logged in the specified file, namely `alert_csv.txt`. The contents of the alert file (`alert_csv.txt`) are as follows:

```
05/13-06:17:08.222534, 4, TCP, stream_tcp, 446, C2S,
145.254.160.237:3372, 65.208.228.223:80, 1:123459991:0, allow
```

The fields listed are the default settings, namely `timestamp`, `pkt_num`, `proto`, `pkt_gen`, `pkt_len`, `dir`, `src_ap`, `dst_ap`, `rule`, and `action`.

Next, we will look at Unified2 format for generating alerts.

Unified2 format

Unified2 format is a binary format. This is one of the most popular formats used by Snort users.

- Command line: `-A unified2`
- Configuration file line: `unified2 = {limit = 1000000}`

Unified2 format has the following configuration parameters:

```
bool unified2.legacy_events = false: generate Snort 2.X style events
for barnyard2 compatibility
int unified2.limit = 0: set maximum size in MB before rollover (0 is
unlimited) { 0:maxSZ }
bool unified2.nostamp = true: append file creation time to name (in
Unix Epoch format)
```

The `unified2.limit` parameter is used to specify the size of the alert file before the rollover operation takes over. The `unified2.legacy_events` parameter can be used for backward compatibility with Snort 2.x versions.

Let's run Snort to use the Unified2 alert format and repeat the same experiment as before.

Exercise

Here's the configuration line in `snort.lua`:

```
unified2 = {limit = 1000000}
```

The command line used for this exercise is as follows:

```
snort3 -c snort.lua -l ~/Downloads/log/ -R ~/Downloads/Snort3/rules/
local.rules --daq dump -Q -r ~/Downloads/PCAPS/http.cap
```

The Unified2 alert was logged in the specified logging directory, namely `~/Downloads/log/`:

```
ls -ltr ~/Downloads/log/
-rw------- 1 auser auser    760 Sep 19 01:12 unified2.log
```

Since Unified2 format is binary, we cannot read it directly. Snort provides a tool called `u2spewfoo`, which is available in the `tools/` directory and can be used to read Unified2 alerts.

The command used for this is as follows:

```
~/Downloads/snort3/build/tools/u2spewfoo/u2spewfoo ~/Downloads/log/
unified2.log
```

Here's the output:

```
(Event)                                        ×
        Snort ID: 0      Event ID: 1     Seconds: 1084443428.222534
        Policy ID:       Context: 0      Inspect: 0      Detect: 0
        Rule 1:123456789:0      Class: 0        Priority: 0
        MPLS Label: 0    VLAN ID: 0      IP Version: 0x44      IP Proto: 6
        Src IP: 145.254.160.237 Port: 3372
        Dst IP: 65.208.228.223  Port: 80
        App Name: none
        Status: allow   Action: pass

(ExtraDataHdr)
        event type: 4    event length: 46

(ExtraData)
        sensor id: 0     event id: 1     event second: 1084443428
        type: 9 datatype: 1     bloblength: 22  HTTP URI: /download.html

(ExtraDataHdr)
        event type: 4    event length: 48

(ExtraData)
        sensor id: 0     event id: 1     event second: 1084443428
        type: 10         datatype: 1     bloblength: 24  HTTP Hostname:
www.ethereal.com

Buffer
        sensor_id: 0     event_id: 1     event_second: 1084443428
        packet_second: 1084443428       packet_microsecond: 222534
        packet_length: 479
[    0] 47 45 54 20 2F 64 6F 77 6E 6C 6F 61 64 2E 68 74  GET /download.ht
[   16] 6D 6C 20 48 54 54 50 2F 31 2E 31 0D 0A 48 6F 73  ml HTTP/1.1..Hos
[   32] 74 3A 20 77 77 77 2E 65 74 68 65 72 65 61 6C 2E  t: www.ethereal.
[   48] 63 6F 6D 0D 0A 55 73 65 72 2D 41 67 65 6E 74 3A  com..User-Agent:
[   64] 20 4D 6F 7A 69 6C 6C 61 2F 35 2E 30 20 28 57 69   Mozilla/5.0 (Wi
[   80] 6E 64 6F 77 73 3B 20 55 3B 20 57 69 6E 64 6F 77  ndows; U; Window
[   96] 73 20 4E 54 20 35 2E 31 3B 20 65 6E 2D 55 53 3B  s NT 5.1; en-US;
[  112] 20 72 76 3A 31 2E 36 29 20 47 65 63 6B 6F 2F 32   rv:1.6) Gecko/2
[  128] 30 30 34 30 31 31 33 0D 0A 41 63 63 65 70 74 3A  0040113..Accept:
[  144] 20 74 65 78 74 2F 78 6D 6C 2C 61 70 70 6C 69 63   text/xml,applic
[  160] 61 74 69 6F 6E 2F 78 6D 6C 2C 61 70 70 6C 69 63  ation/xml,applic
[  176] 61 74 69 6F 6E 2F 78 68 74 6D 6C 2B 78 6D 6C 2C  ation/xhtml+xml,
[  192] 74 65 78 74 2F 68 74 6D 6C 3B 71 3D 30 2E 39 2C  text/html;q=0.9,
[  208] 74 65 78 74 2F 70 6C 61 69 6E 3B 71 3D 30 2E 38  text/plain;q=0.8
[  224] 2C 69 6D 61 67 65 2F 70 6E 67 2C 69 6D 61 67 65  ,image/png,image
[  240] 2F 6A 70 65 67 2C 69 6D 61 67 65 2F 67 69 66 3B  /jpeg,image/gif;
[  256] 71 3D 30 2E 32 2C 2A 2F 2A 3B 71 3D 30 2E 31 0D  q=0.2,*/*;q=0.1.
[  272] 0A 41 63 63 65 70 74 2D 4C 61 6E 67 75 61 67 65  .Accept-Language
[  288] 3A 20 65 6E 2D 75 73 2C 65 6E 3B 71 3D 30 2E 35  : en-us,en;q=0.5
[  304] 0D 0A 41 63 63 65 70 74 2D 45 6E 63 6F 64 69 6E  ..Accept-Encodin
[  320] 67 3A 20 67 7A 69 70 2C 64 65 66 6C 61 74 65 0D  g: gzip,deflate.
[  336] 0A 41 63 63 65 70 74 2D 43 68 61 72 73 65 74 3A  .Accept-Charset:
[  352] 20 49 53 4F 2D 38 38 35 39 2D 31 2C 75 74 66 2D   ISO-8859-1,utf-
[  368] 38 3B 71 3D 30 2E 37 2C 2A 3B 71 3D 30 2E 37 0D  8;q=0.7,*;q=0.7.
[  384] 0A 4B 65 65 70 2D 41 6C 69 76 65 3A 20 33 30 30  .Keep-Alive: 300
[  400] 0D 0A 43 6F 6E 6E 65 63 74 69 6F 6E 3A 20 6B 65  ..Connection: ke
[  416] 65 70 2D 61 6C 69 76 65 0D 0A 52 65 66 65 72 65  ep-alive..Refere
[  432] 72 3A 20 68 74 74 70 3A 2F 2F 77 77 77 2E 65 74  r: http://www.et
[  448] 68 65 72 65 61 6C 2E 63 6F 6D 2F 64 65 76 65 6C  hereal.com/devel
[  464] 6F 70 6D 65 6E 74 2E 68 74 6D 6C 0D 0A 0D 0A     opment.html....
```

Figure 14.2 – Using u2spewfoo to read a Unified2 alert

Unified2 is one of the most widely used alerting formats for Snort. Next, let's discuss Alert Fast format.

Alert Fast format

This is a format where the alert is very brief. Unlike CSV format, we cannot specify which fields to log as part of the alert. This is usually used for quick testing or verification.

- Command line: `-A alert_fast`
- Configuration file line: `alert_fast = {file = true, limit = 1000000}`

The configuration parameters that are available for this format are as follows:

```
bool alert_fast.file = false: output to alert_fast.txt instead of
stdout
bool alert_fast.packet = false: output packet dump with alert
int alert_fast.limit = 0: set maximum size in MB before rollover (0 is
unlimited) { 0:maxSZ }
```

The `alert_fast.limit` parameter is used to specify the size of the alert file before the rollover operation takes over. The `alert_fast.file` parameter specifies where the alert should be logged; `true` causes the alert to be logged to the file named `alert_fast.txt`, whereas `false` causes the alert to be logged to `stdout`. The `alert_fast.packet` parameter, if set to `true`, will cause the packet bytes to be logged along with the alert.

In our experiment, we specify the file parameter and the limit.

Exercise

Let's look at the relevant configuration line in `snort.lua`:

```
alert_fast = {file = true, limit = 1000000}
```

The command line used for this exercise is as follows:

```
snort3 -c snort.lua -l ~/Downloads/log/ -R ~/Downloads/Snort3/rules/
local.rules --daq dump -Q -r ~/Downloads/PCAPS/http.cap
```

The alert file was logged in the specified logging directory, namely `~/Downloads/log/`.

The contents of the alert file, `alert_fast.txt`, are as follows:

```
05/13-06:17:08.222534 [**] [1:123459991:0] "Download HTML Rule" [**]
[Priority: 0] {TCP} 145.254.160.237:3372 -> 65.208.228.223:80
```

Alert Fast format is useful when a lightweight alerting functionality is needed, especially when troubleshooting signatures or other modules. Next, we will discuss Alert Full format.

Alert Full format

Alert Full format provides more details than Alert Fast format, and it also provides a full PCAP as part of the alert.

- Command line: `-A alert_full`
- Configuration file line: `alert_full = {file = true, limit = 1000000}`

This option has the following configuration parameters:

```
bool alert_full.file = false: output to alert_full.txt instead of
stdout
int alert_full.limit = 0: set maximum size in MB before rollover (0 is
unlimited) { 0:maxSZ }
```

Exercise

The configuration line in `snort.lua` to enable the `alert_full` module is as follows:

```
alert_full = {file = true, limit = 1000000}
```

The same experiment is repeated but with the `alert_full` module for alerting. The command line used with this exercise is as follows:

```
snort3 -c snort.lua -l ~/Downloads/log/ -R ~/Downloads/Snort3/rules/
local.rules --daq dump -Q -r ~/Downloads/PCAPS/http.cap
```

The alert file was logged in the specified logging directory, namely `~/Downloads/log/`.

The contents of the alert file, `alert_full.txt`, are as follows:

```
[**] [1:123459991:0] Download HTML Rule [**]
[Priority: 2]
05/13-05:17:10.295515 145.254.160.237:3371 -> 216.239.59.99:80
TCP TTL:55 TOS:0x10 ID:34104 IpLen:20 DgmLen:761
***A**** Seq: 0x36C21E28 Ack: 0x2E6B5384 Win: 0x7AE4 TcpLen: 20
```

The `alert_full` module is useful when we require more details in the alert than we get with the `alert_fast` option. Next, we will discuss JSON format alerting.

JSON format

This option provides alerts in JSON format. The following configuration parameters are available for the `alert_json` option:

```
bool alert_json.file = false: output to alert_json.txt instead of
stdout
multi alert_json.fields = 'timestamp pkt_num proto pkt_gen pkt_len
dir src_ap dst_ap rule action': selected fields will be output in
given order left to right { action | class | b64_data | client_bytes |
client_pkts | dir | dst_addr | dst_ap | dst_port | eth_dst | eth_len
| eth_src | eth_type | flowstart_time | geneve_vni | gid | icmp_code |
icmp_id | icmp_seq | icmp_type | iface | ip_id | ip_len | msg | mpls |
pkt_gen | pkt_len | pkt_num | priority | proto | rev | rule | seconds
| server_bytes | server_pkts | service | sgt| sid | src_addr | src_ap
| src_port | target | tcp_ack | tcp_flags | tcp_len | tcp_seq | tcp_
win | timestamp | tos | ttl | udp_len | vlan }
int alert_json.limit = 0: set maximum size in MB before rollover (0 is
unlimited) { 0:maxSZ }
string alert_json.separator = ', ': separate fields with this
character sequence
```

Like CSV format, JSON format also provides a `config` parameter to specify which fields should be logged as part of the alert.

Exercise

Here's the configuration line in `snort.lua`:

```
alert_json = {file = true, limit = 1000000}
```

The command line used for this exercise is as follows:

```
snort3 -c snort.lua -l ~/Downloads/log/ -R ~/Downloads/Snort3/rules/
local.rules --daq dump -Q -r ~/Downloads/PCAPS/http.cap
```

The alert file was logged in the specified logging directory, namely `~/Downloads/log/`.

The contents of the alert file, `alert_json.txt`, are as follows:

```
{ "timestamp" : "05/13-06:17:08.222534", "pkt_num" : 4, "proto" :
"TCP", "pkt_gen" : "stream_tcp", "pkt_len" : 446, "dir" : "C2S", "src_
ap" : "145.254.160.237:3372", "dst_ap" : "65.208.228.223:80", "rule" :
"1:123459991:0", "action" : "allow" }
```

The JSON alert format lists the names of the fields. The `src_ap` field lists the source IP address and port, and similarly `dst_ip` lists the destination IP address and port. The `dir` key lists the direction of the traffic, and `C2S` stands for Client to Server.

In this section, we looked at the various formats of alerts created by the alert subsystem.

Summary

In this chapter, we discussed the alert subsystem of Snort. We discussed the process that happens when there is a successful match for a signature. We discussed the role of the alert subsystem, that is, to create an alert when there is a successful identification of a malicious packet or session.

We looked at the various alert formats and looked at a few formats in detail. In the next chapter, we will explore OpenAppID.

15
OpenAppID

In the previous chapters, we learned about the different modules of Snort 3 IDS/IPS, which essentially performs in-depth analysis of network traffic in order to detect malicious behavior and exploit attempts. Toward this goal, the users would maintain a set of IDS/IPS signatures that work in conjunction with Snort modules to detect and stop bad traffic.

In this chapter, we have a different use case that is practically useful. Network administrators and/or policymakers of organizations often like to limit and/or control the use of certain applications within the environment. For example, the network admin or controller may want to limit access (block access) to iTunes traffic. Note that this is not a security problem; rather, it is a policy issue. Historically, Snort rules were written to detect traffic of a particular application and thus alert and block it. These rules were grouped as `policy.rules`.

The OpenAppID feature is the answer to this use case. The OpenAppID feature enables the creation of Lua based scripts that can identify applications. These, in conjunction with Snort rules (using a special rule option called `appid`), can be used to alert/log/block applications.

In this chapter, we will discuss the OpenAppID feature and the relevant inspector modules, and their configuration, covered under the following headings:

- The OpenAppID feature
- Design and architecture

The OpenAppID feature

The term AppID in the word OpenAppID indicates what the module does, namely **application identification**. Application identification is a key feature of **next-generation firewalls** (**NGFWs**). This feature enables Snort to perform the same level of analysis as NGFWs in addition to the IPS/IDS functionality.

Awareness of the application that is associated with network traffic is valuable information. This enables the system to control and enforce policies; it also adds more context to the remaining network traffic analysis and rule matching.

There is a main difference between the OpenAppID feature and most other features of Snort. While other features are geared toward detecting badness and stopping attacks, the OpenAppID feature is, by design, not aimed at detecting attacks and exploits. Rather, this feature is designed to detect common applications so that the network administrators can detect usage and enforce policies (for example, an organization may not allow Facebook usage).

The OpenAppID feature can be considered as a *layer* within Snort for application identification. The OpenAppID module aims to detect the following aspects of each network flow:

- Service

- Client

- Application

For example, if an end host browser is used to access iTunes or Facebook, OpenAppID may detect the service as HTTP, the client as Firefox, and the application as iTunes or Facebook. This detection analysis is done on a per-flow basis. Therefore, basic flow tracking and TCP/UDP session tracking are necessary and a dependency for the OpenAppID feature.

The detection feature of OpenAppID works in combination with the `appid` rule option feature of the module. This rule option can be used in Snort signatures to match against specific applications or services – with the aim of creating policy violation rules and/or policy enforcement rules.

Design and architecture

The OpenAppID feature has a few components or parts, namely, the following:

- **Detectors**: These components have the logic for the application detection. These are written in the Lua scripting language. These are packaged as a separate package and not as part of the Snort package (similar to how Snort rules are a separate package). This is done so that the detectors can be developed, tested, and released separately from the Snort package.

- **Rules using the application identification information**: These are the Snort rules that utilize the `appid` rule option.

- **The OpenAppID inspector**: This fits into the Snort architecture as a network inspector module.

Let us look at these components a bit more closely in the next section.

Detectors

The application detectors are the key components as far as OpenAppID is concerned. This is the meat of the functionality. This logic is mainly written in Lua scripts. These inspect the client-side and server-side traffic for certain conditions to complete the detection. In certain cases, the detection is done purely based on ports. This is simple and straightforward (possibly error-prone); for example, a rule might map the traffic to an HTTP proxy when the port involved is TCP/3128. In other cases, the detection is done based on specific patterns that it tries to match in the client and server-side packets. Such detection may happen from inspecting a few bytes of the initial traffic on a connection, or it may take a few packets. Any relevant state with respect to the application discovery is stored as part of the flow that is tracked by Snort.

Open Detector Package

The OpenAppID detectors are available to be downloaded on the Snort website at the following link: https://www.snort.org/downloads#openappid. The package is called **Open Detector Package** (**ODP**). The package contains two types of detectors, namely application detectors and port detectors.

The port detectors are Lua scripts that have a simple port-based logic; they map the session to a particular application if the port used is a certain value. The application detectors are based on content, usually identified by the very first packet. These Lua scripts specify the pattern that should be seen and matched in the client or server packet to be detected as a certain application.

From a packaging and maintenance perspective, the OpenAppID detectors are similar to Snort rules. There is a large community that contributes and participates in the creation of the detectors. Like Snort rules are packaged separately from the Snort engine package, the **Open Detector Package** (**ODP**) is also separate.

The Open Detector Package can be installed in any directory location. However, that location needs to be specified in the appId Inspector configuration. The package has the following contents:

- **The Lua-based detectors** (both port-based and application-based) : These are created for the detection of specific applications. Among these detectors there are port-based detectors that base their detection logic on the layer 4 port numbers. The other type of detectors base their logic on application protocol syntax and other details. These are programmed using the Lua language.

- **Configuration files**: These include the appMapping.data file that contains data that maps the application identifier (number) to the application name.

  The appMapping.data file contains the lines that follow:

  ```
  2       3COM-TSMUX       0      0      0      ~
          3com-tsmux
  4       914CG            0      0      0      ~
          914cg
  5        ACA Services    0      0      0      ~
          aca_services
  ```

  The first column is for the application ID, and the last column is for the application name as referenced by the rules.

- **Library files**: The package also contains a `lib/` directory with the `DetectorCommon.lua` file. This is an important file that is included in every Lua-based detector.

In the next section, we will discuss the OpenAppID inspector.

The inspector

The OpenAppID inspector is grouped with the network inspectors (in the Snort code base). It is the role of this inspector to identify the application associated with both sides of every connection (client- and server-side). Since each direction of a network connection is treated as a flow, it is better to associate the identification with each flow.

The OpenAppID inspector inspects the packets of each flow, and with the help of the detectors, it identifies the associated application. The inspector depends on the flow object for its working; the state associated with application discovery is stored in the flow object.

Each detector registers a set of ports and a set of fast patterns during its registration phase. As the OpenAppID inspector processes the packets for any flow, it checks the ports and the fast patterns and invokes the appropriate `validate` function of the detector when there is a match. The `validate` function makes one of the following decisions: match, no match, or pending decision.

Inspector configuration

The configuration for OpenAppID is done in `snort.lua` (or the relevant Lua configuration file). For example, a sample configuration for the inspector is as follows:

```
appid =
{
    app_detector_dir = '/usr/local/snort/odp'
    log_stats = true
}
```

As we discussed in the section about Open Detector Package, the OpenAppID package is downloaded and installed separately from the Snort package. It is using the `app_detector_dir` parameter that we tell Snort where to find the detectors.

The OpenAppID inspector also keeps several statistics of the identified application. The next parameter is used to enable the logging of the statistics.

The `snort --help-module` command can be used to list out all the relevant parameters that can be configured for OpenAppID. The output of the command (`snort --help-module appid`) is shown further. Note that the output for this command has three sections, namely `Configuration`, `Commands`, and `Peg Counts`. We will discuss only the `Configuration` section here:

```
snort --help-module appid
Configuration:
```

```
int appid.memcap = 1048576: max size of the service cache before we
start pruning the cache { 1024:maxSZ }
bool appid.log_stats = false: enable logging of appid statistics
int appid.app_stats_period = 300: time period for collecting and
logging appid statistics { 1:max32 }
int appid.app_stats_rollover_size = 20971520: max file size for appid
stats before rolling over the log file { 0:max32 }
string appid.app_detector_dir: directory to load appid detectors from
bool appid.list_odp_detectors = false: enable logging of odp detectors
statistics
bool appid.log_all_sessions = false: enable logging of all appid
sessions
bool appid.enable_rna_filter = false: monitor only the networks
specified in rna configuration
string appid.rna_conf_path: path to rna configuration file
```

As can be noted, the two parameters that we listed in the sample Lua configuration are also part of the listed parameters (app_detector_dir and log_stats).

There are three parameters that control the collection and logging of the statistics. The log_stats parameter is used to enable the logging of the statistics, and the app_stats_period parameter can be used to specify at what frequency the values are logged. A value of 300 for app_stats_period will cause the module to log the statistics every 5 minutes. Finally, the app_stats_rollover_size parameter is used to specify the file size at which the file rolls over (to a new file).

The list_odp_detectors parameter is used to enable the logging of statistics maintained by the detectors as well. The log_all_sessions parameter is a control to enable the logging of all the sessions analyzed by the module.

The module also supports a method to have the application identification done only for a subset of traffic. If enable_rna_filter is enabled, then only the networks specified in the **Realtime Network Awareness** (**RNA**) configuration file will be monitored.

In the next section, let us discuss how the rules take advantage of the application identification that is performed by the OpenAddID inspector.

The rules

The Snort rules take advantage of the work done by the OpenAppID feature by making use of the related rule option called appids. When we use this rule option in a Snort rule, that rule will only trigger when the identified application is the specified value. Let's consider an example rule as given here:

```
alert tcp any any -> any any (msg:"Kismet Traffic"; appids:"kismet";
sid:12345; )
```

Notice that the `appids` rule option specifies `"kismet"`. The name of the application, in this case, is `"kismet"`. The name of the application that is used in the rule along with the `appids` keyword must be consistent with the application name specified in the `appMapping.data` file in the Open Detector Package.

The preceding rule will only match flows that have been identified as `"kismet"`. The OpenAppID inspector, with the help of the corresponding detector (the `service_Kismet.lua` detector file installed as part of the Open Detector Package) identifies the application associated with the session. If the rule is changed from `alert` to `drop`, then this rule can be used to stop Kismet traffic. It would be good to bring together all the information we have learned using an exercise. Let's do that in the next section.

Exercise

Let's go through an exercise combining all the information we have discussed. In this *Exercise* section, we will discuss the use case of the Kismet protocol and the associated traffic flows. The Kismet protocol is used by the Kismet client and Kismet server, which are a GUI program and a sniffer machine. We will see how the detector detects the Kismet protocol, and how a rule can take advantage of the identification.

Kismet detection

For the detection of the Kismet application (traffic), we could use the commonly used port, as well as a byte sequence in the network traffic that would identify Kismet traffic. The standard port associated with Kismet is `2501/TCP`. In addition, the server responds at the start of the flow with the `"KISMET: "` bytes, as can be seen in the following figure:

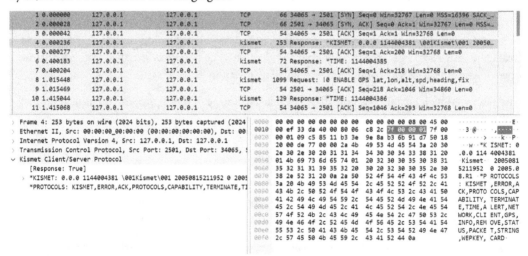

Figure 15.1 – Network traffic for the Kismet protocol

Now let us look at some sections of OpenAppID Detector for Kismet. As mentioned before, the filename is `service_Kismet.lua`:

```
gServiceId = 20140
gServiceName = 'Kismet'
gSfAppIdKismet = 1451
gPorts = {
    {DC.ipproto.tcp, 2501},
}
gPatterns = {
    server_resp = {'KISMET:', 1, gSfAppIdKismet},
}
gFastPatterns = {
    {DC.ipproto.tcp, gPatterns.server_resp},
}
```

The preceding snippet shows the section of the `service_Kismet.lua` detector where it specifies the port that is relevant to the Kismet traffic, namely TCP/2501. In addition, we see the fast pattern that is used to identify the Kismet application, namely 'KISMET:'.

When the inspector module sees the traffic on TCP/2501 and the packet contains the 'KISMET:' string, it invokes the validator function within the detector module. Once the validator successfully validates the session as Kismet, then the session/flow is identified as Kismet. Thereafter, a rule with the rule option – appids:'kismet' – will succeed for this traffic flow. This is what we will discuss in the next section.

Snort rule for the exercise

We need a Snort rule that checks whether the session is detected as Kismet. The following Snort rule does that:

```
alert tcp any any -> any any (msg:"Kismet Traffic"; appids:"kismet";
sid:12345; )
```

Snort execution

We install the OpenAppID package in a specific directory. In this example, the package is installed under /home/kali/snort/ODP. We take advantage of a default file that comes with the OpenAppID package. Make sure we have the Kismet app detector under the lua directory in the installed OpenAppID directory:

```
ls /home/kali/snort-source-files/ODP/odp/lua| grep -i kismet
service_Kismet.lua
```

We edit the snort.lua file to enable the OpenAppID inspector with the following line:

```
appid =
{
    app_detector_dir = '/home/kali/snort/ODP'
    log_stats = true
}
```

Now, let's run Snort as follows:

```
snort -c ~/snort-source-files/snort3/lua/snort.lua -r ~/Downloads/
kismet-client-server-dump-1.pcap -R ~/Rules/local.rules -k none -l ~/
Log
```

The alert file is stored under the ~/Log directory:

```
04/02-14:59:45.285561 [**] [1:12345:0] "Kismet Traffic"
[**] [Priority: 0] [AppID: Kismet] {TCP} 127.0.0.1:2501 ->
127.0.0.1:34065
04/02-14:59:45.285602 [**] [1:12345:0] "Kismet Traffic" [**]
[Priority: 0] [AppID: Kismet] {TCP} 127.0.0.1:34065 ->
127.0.0.1:2501
04/02-14:59:45.685508 [**] [1:12345:0] "Kismet Traffic"
[**] [Priority: 0] [AppID: Kismet] {TCP} 127.0.0.1:2501 ->
127.0.0.1:34065
04/02-14:59:45.685529 [**] [1:12345:0] "Kismet Traffic" [**]
[Priority: 0] [AppID: Kismet] {TCP} 127.0.0.1:34065 ->
127.0.0.1:2501
```

The rule would work irrespective of which TCP port is used by the Kismet application. OpenAppID will detect the application, and the rule will be applied for that traffic.

In this exercise, we connected all that we learned in this chapter. We saw how the OpenAppID detector detects a Kismet session. Such detection can happen even though the protocol may be used on a random ephemeral port. Thereafter, we saw how a rule can check whether the session is detected as Kismet, and then create an alert. This is the alert that Snort produced as part of the exercise.

Summary

In this chapter, we learned about the application identification feature in Snort, which is known as the OpenAppID feature. We learned about the feature and its use. We discussed the OpenAppID inspector module and its configuration settings. The rule options that makes use of the application identification feature was also discussed.

The chapter also discussed the OpenAppID detector package, which must be downloaded and installed separately from Snort. This package is maintained separately and enjoys large community support. In the next and final chapter, we will discuss some miscellaneous topics, including how to troubleshoot a Snort crash, and how to migrate a Snort 2 configuration to Snort 3.

16
Miscellaneous Topics on Snort 3

This chapter covers a variety of different topics that are relevant to Snort 3. We will cover topics such as Snort troubleshooting and debugging, Snort 2 to Snort 3 migration challenges, and more.

Snort 2 has been in use for more than a decade, and there have been millions of downloads of the program over these years. Perhaps you have a working configuration and setup that you would like to migrate as part of upgrading to Snort 3.

Debugging and troubleshooting are key phases of any software, and that is expected to be the case for Snort 3 as well. Such debugging or troubleshooting could be related to testing a signature, or it could be related to verifying the correct functioning of a certain module, or it could be related to performance.

In this chapter, we will discuss these topics and discuss examples to equip you with practical knowledge that will be useful.

The topics we discuss in this chapter are as follows:

- Snort 2 to Snort 3 migration
- Troubleshooting Snort 3

Snort 2 to Snort 3 migration

Let's say you have an IDS or IPS system using Snort 2. You also have a working Snort 2 configuration and a set of rules that you have built over a considerable period of time. Now, you wish to migrate to Snort 3. There are a few challenges. Firstly, Snort 3 syntax has changed and is not backward compatible with Snort 2. Secondly, Snort 3 uses a Lua-based configuration, and we cannot use the existing Snort 2 configuration as is. In this section, let's discuss the topic of migrating from Snort 2 to Snort 3.

Migrating the rules

Snort 3 extends the detection and signature matching capabilities of the Snort IDS/IPS platform. Along with this, some of the syntax rules have also changed and it does not provide backward compatibility to the Snort 2 equivalent. To use any Snort 2 signatures with Snort 3, they need to be converted to their Snort 3 versions.

Although the Snort team recommends rewriting the signatures manually, taking into account all the new features and capabilities, they have provided a tool (script named **snort2lua**) that automatically converts a Snort 2 signature to its Snort 3 equivalent.

Snort2lua

Snort2lua is a tool that is shipped with the Snort 3 package. If you are downloading the source, you will find the code in the `tools/` directory (for example, `~/snort3-3.1.47.0/build/tools/snort2lua`).

The tool covers the conversion of Snort 2 configuration and Snort 2 rules to Snort 3 versions. In this section, we will only discuss the conversion of rules. The command to convert rules is as follows:

```
./snort2lua -c snort2rules.rules -r snort3rules.rules
```

This command takes `snort2rules.rules` as the input file, which contains rules that are compatible with Snort 2, and writes the translated Snort 3 version rules into the `snort3rules.rules` file.

Relevant Snort2lua command-line options are as follows:

- `-?`: Shows usage. This option lists out all the possible command line arguments that the tool takes.

- `-h`: This gives an overview of Snort2lua.

- `-c <snort_conf>`: The Snort configuration filename to be converted is specified using this option.

- `-r <rule_file>`: This outputs any converted rule to `<rule_file>`.

When there are any issues in converting the rule, the tool logs the rule in a file called `snort.rej`.

The Snort2lua tool converts all the Snort 2 rules to a compatible Snort 3 version without any errors, in most cases. However, it should be noted that the converted rules are not optimized for Snort 3. They work but are not the best version possible. Also, if the rule uses obsolete options such as `threshold`, the Snort2lua tool may fail to convert the rule as expected.

To test the tool, we can pull the Snort 2.9 rules from **EmergingThreats** (**ET**). You can find the ET rules at `rules.emergingthreats.net/open/snort-2.9.0/emerging-all.rules`. From this set, we randomly selected ten rules and converted them using Snort2lua. There were no errors. Let's discuss two of these rules.

The first rule detects the **command and control** traffic for a Trojan called Oceanlotus. The Snort 2.9 version of the rule is provided here, and it should be noted that it is a rule that uses HTTP-related keywords and HTTP buffers.

Let's see how Snort2lua converts this rule:

```
alert tcp $HOME_NET any -> $EXTERNAL_NET $HTTP_PORTS (msg:"ET
TROJAN Suspected APT32/Oceanlotus Maldoc CnC"; flow:established,to_
server; content:"GET"; http_method; content:".png?"; http_uri;
content:"=e010000127"; distance:0; fast_pattern; http_uri;
content:".exe|3b|"; nocase; http_uri; pcre:"/^[^\r\n]+\.exe(?:\
x3b)?$/Ui"; reference:md5,e2511f009b1ef8843e527f765fd875a7;
reference:md5,cc2027319a878ee18550e35d9b522706; reference:url,twitter.
com/HONKONE_K/status/1290511333343993856; classtype:trojan-activity;
sid:2030652; rev:2; metadata:attack_target Client_Endpoint, created_
at 2020_08_05, deployment Perimeter, deployment SSLDecrypt, former_
category MALWARE, malware_family APT32, malware_family OceanLotus,
performance_impact Low, signature_severity Major, updated_at
2020_08_05;)
```

This rule was converted by Snort2lua, as shown in the following command block. We immediately notice the use of sticky buffers in the Snort3 version. In addition, we also notice how content modifiers such as distance are used in conjunction with the content keyword instead of as separate terms:

```
alert tcp $HOME_NET any -> $EXTERNAL_NET $HTTP_PORTS (
msg:"ET TROJAN Suspected APT32/Oceanlotus Maldoc CnC";
flow:established,to_server; http_method; content:"GET"; http_
uri; content:".png?"; content:"=e010000127",distance 0,fast_
pattern; content:".exe|3b|",nocase; pcre:"/^[^\r\n]+\.exe(?:\
x3b)?$/i"; reference:md5,e2511f009b1ef8843e527f765fd875a7;
reference:md5,cc2027319a878ee18550e35d9b522706; reference:url,twitter.
com/HONKONE_K/status/1290511333343993856; classtype:trojan-activity;
sid:2030652; rev:2; metadata:attack_target Client_Endpoint,created_at
2020_08_05,deployment Perimeter,deployment SSLDecrypt,former_category
MALWARE,malware_family APT32,malware_family OceanLotus,performance_
impact Low,signature_severity Major,updated_at 2020_08_05; )
```

Let's take a look at one more signature. The next rule is a simpler rule and detects malicious traffic related to a malware that is named W32/Coced.PasswordStealer. The rule is based on a unique User-Agent header option in HTTP. Let's see how Snort2lua converts this rule:

```
alert tcp $HOME_NET any -> $EXTERNAL_NET $HTTP_PORTS (msg:"ET TROJAN
W32/Coced.PasswordStealer User-Agent 5.0"; flow:established,to_
server; content:"User-Agent|3A 20|5.0|0D 0A|"; http_header;
reference:md5,24e937b9f3fd6a04dde46a2bc75d4b18; classtype:trojan-
activity; sid:2014344; rev:1; metadata:created_at 2012_03_09, updated_
at 2012_03_09;)
```

This rule was converted by Snort2lua, as shown in the following command block. This is a relatively straightforward conversion. Again, we notice the use of sticky buffers in the Snort3 variant:

```
alert tcp $HOME_NET any -> $EXTERNAL_NET $HTTP_PORTS ( msg:"ET TROJAN
W32/Coced.PasswordStealer User-Agent 5.0"; flow:established,to_
server; http_header; content:"User-Agent|3A 20|5.0|0D 0A|";
reference:md5,24e937b9f3fd6a04dde46a2bc75d4b18; classtype:trojan-
activity; sid:2014344; rev:1; metadata:created_at 2012_03_09,updated_
at 2012_03_09; )
```

Overall, the Snort2lua tool is a great tool for converting the rules we have written for Snort 2 to their Snort 3 equivalent.

In the next section, let's see how Snort2lua is useful for migrating Snort configurations.

Migrating configurations

Snort 3 configurations use the Lua language. Therefore, when migrating existing Snort 2.9 configurations, the migrated `config` file should be in Lua. Let's see how well Snort2lua manages to convert the default Snort 2.9 configuration. We took the default `snort.conf` from Snort 2.9 and fed it into Snort2lua as follows:

```
./snort2lua -c snort.conf -o snort3.lua
```

Snort variables are a part of the Snort configuration. There are different types of variables, such as `ipvar` and `portvar`. The Snort2lua tool seamlessly converts these variables as we shall see in the next section.

Snort variables

Snort2lua converted all the `ipvar`, `portvar`, and other variables as follows:

Snort 2.9	Snort 3
`ipvar HOME_NET any`	`HOME_NET = 'any'`
`ipvar EXTERNAL_NET any`	`EXTERNAL_NET = 'any'`
`ipvar DNS_SERVERS $HOME_NET`	`DNS_SERVERS = HOME_NET`
`portvar HTTP_PORTS [80,3128,8000,8080]`	`HTTP_PORTS = [[ 80 3128 8000 8080 ]]`
`var RULE_PATH ../rules`	`RULE_PATH = '../rules'`
`var SO_RULE_PATH ../so_rules`	`SO_RULE_PATH = '../so_rules'`

Table 16.1 – Conversion of Snort variables by Snort2lua

The Snort2lua tool converts all the configuration variables of types ipvar and portvar into LUA variables as shown in *Table 16.1*.

In some cases, the migration of configurations from Snort 2.9 to Snort 3 is not straightforward. In the next section, let's look at how Snort2lua handles configuration options that are no longer supported in Snort 3.

Obsolete configuration options

There are configuration options in Snort 2.9 that are no longer supported in Snort 3. In other cases, a configuration option in Snort 2.9 may be called a different name in Snort 3. When Snort2lua is used, it translates these config options as needed and adds a comment regarding the translation.

Let's take the example of a configuration related to the detection engine. In Snort 2.9, this is specified using the `config detection` option, as follows:

```
config detection: search-method ac-split search-optimize max-pattern-
len 20
```

This was translated as follows by Snort2lua:

```
search_engine =
{
    search_method = 'ac_full',
    max_pattern_len = 20,
    split_any_any = true,
    --This table was previously 'config detection: ...
    --option change: 'ac-split' --> 'ac_full'
    --option change: 'ac-split' --> 'split_any_any'
    --option change: 'max-pattern-len' --> 'max_pattern_len'
    --option change: 'search-method' --> 'search_method'
    --option deleted: 'search-optimize is always true'
}
```

The comments left by Snort2lua are very useful. They say that `'ac-split'` in Snort 2.9 has been translated to `'ac_full'` in Snort 3. Similarly, in the case of `'search-optimize'` setting in Snort 2.9, this option is no longer supported in Snort 3 (the optimize feature is always enabled).

Now, let's take a look at how the Snort 2.9 preprocessor configuration is translated to Snort 3. We will look at the `normalize` preprocessors first.

The Snort 2.9 configuration had the following settings:

```
preprocessor normalize_ip4
preprocessor normalize_tcp: ips ecn stream
preprocessor normalize_icmp4
```

```
preprocessor normalize_ip6
preprocessor normalize_icmp6
```

This was correctly translated by Snort2lua as follows:

```
normalizer =
{
    icmp4 = true,
    ip6 = true,
    icmp6 = true,
    ip4 =
    {
    },
    tcp =
    {
        ips = true,
        ecn = 'stream',
    },
    --option change: 'preprocessor normalize_icmp4' --> 'icmp4 =
<bool>'
    --option change: 'preprocessor normalize_icmp6' --> 'icmp6 =
<bool>'
    --option change: 'preprocessor normalize_ip6' --> 'ip6 = <bool>'
}
```

The different preprocessors are replaced by the normalizer inspector. The various settings pertaining to ip, tcp, and udp were translated to the corresponding settings in the normalizer inspector configuration.

Specifically, in the case of `normalize_tcp`, the configuration in Snort 2.9 stated that ips mode was enabled:

```
normalize_tcp: ips ecn stream
```

As we can see, in the Snort 3 settings, this is correctly translated and maintained:

```
tcp =
{
    ips = true,
    ecn = 'stream',
},
```

After using Snort2lua to migrate the configuration, close inspection is required to make sure the settings are what we need.

In addition, the newly created `snort.lua` configuration file must be tested in a controlled way to ensure correct working.

In the next section, let's discuss a few scenarios where we may need to do Snort debugging or troubleshooting.

Troubleshooting Snort 3

After you have migrated the configuration (from Snort 2 to Snort 3) and Snort 3 is running, there are various scenarios where the system is not working as you expected it to. Does the Snort rule you wrote work as expected? What to do if Snort crashes? Where do we reach out for support when needed? Let's discuss a few of these topics in this section.

Why is the Snort rule for XYZ not alerting?

This is a very common question that is often asked. We have a Snort signature that is meant to detect an attack or some malicious traffic. Why is it not alerting when the IDS is inspecting bad traffic?

We will keep this discussion simple and address some of the common reasons:

1. **Get the packet capture**: To debug and troubleshoot the situation, the best way is to get the traffic under consideration as a **packet capture** (**pcap**) file. This will enable us to have Snort analyze the traffic in a controlled fashion, and also repeat the test as many times as needed.

   Once the `pcap` file is obtained, manually analyze the traffic using **tcpdump** or **Wireshark** and verify that the use case is accurate (i.e., that the traffic is indeed malicious and expected to match the signature).

2. Does the rule header specification (IP, port, and protocol) match the traffic's IP, port, and protocol?

   I.   For example, does the rule say this?

   ```
   alert tcp $HOME_NET any -> $EXTERNAL_NET $HTTP_PORTS
   ```

   If so, match this with the traffic packet we expect to hit the signature, and verify the following:

   i.   The protocol is `TCP`.

   ii. The source IP address matches the `HOME_NET` setting in the Snort configuration.

   iii. The destination IP address matches the `EXTERNAL_NET` setting in the Snort configuration.

   iv. The destination port matches the specified `HTTP_PORTS` list in the Snort configuration.

   II.   If these verifications pass, then we move on to the details of the rules. In the case of Snort rules, once the multi-pattern stage is done, the rule is evaluated from left to right. So, in order to continue our check, we have to look at the rule options one by one from left to right and check whether it is expected to succeed.

Let's look at an example:

```
alert tcp $HOME_NET any -> $EXTERNAL_NET $HTTP_PORTS (
msg:"ET TROJAN Suspected APT32/Oceanlotus Maldoc CnC";
flow:established,to_server; http_method; content:"GET"; http_
uri; content:".png?"; content:"=e010000127",distance 0,fast_
pattern; content:".exe|3b|",nocase; pcre:"/^[^\r\n]+\.exe(?:\
x3b)?$/i"; sid:2030652; rev:2; )
```

The msg option is just used in alerting. The first thing we check is the flow option, which specifies that the session has to be established and the direction of the traffic is "to server". Here, we check the pcap, verify that the TCP three-way handshake is present, and verify that the packet of interest is indeed going to the server (refer to the client and server definitions in TCP).

III.   Subsequently, we verify that the traffic is HTTP, and the HTTP method used is GET. In this fashion, we evaluate the rule against the traffic manually by close inspection. Such analysis often identifies the issue for the rule not triggering , such as a mismatch between the traffic and what the rule matches.

3.   At this point, if we still believe that the rule is expected to match the traffic, then it may be useful to check how the alert module is configured:

I.   Check whether the alert module is enabled.

II.   Verify the location where the alert is configured to be written.

Next, let's look at another scenario.

Snort is crashing!

Snort crashes are extremely rare, but they do happen. There are various reasons why Snort could crash. One of the causes of Snort crashes is a **segmentation fault error**. In any case, there are a few steps that need to be taken to identify the issue and make progress:

1.   Check whether the issue is repeatable.

    Enable the system to dump the core if the Snort crashes. We do this using the following setting (on Linux):

```
ulimit -c unlimited
```

    If we get a core dump file, we can use gdb to investigate the core file and get information on where the issue is. We may need Snort to be compiled with the –gdb option to get such details. However, if the problem is repeatable, then we could recompile Snort using the –gdb option and recreate the core dump.

2. Report the issue to the Snort developers. We can file a bug at `https://github.com/snort3/snort3/issues` as well. When we create a bug or report an issue, we must ensure that we provide the following:

   - The Snort version

   - The command line that was used to run Snort

   - The Snort configuration

   - If there is a `pcap` that can recreate the problem, then provide it

Help! Got support?

There are scenarios where we need some help and support regarding Snort 3. The following mailing lists are useful for this purpose:

- **Snort Users**: This mailing list contains general information and discussions about Snort. This is useful for beginners and advanced users.

- **Snort Sigs**: This mailing list is for discussions pertaining to Snort signatures or rules.

- **Snort Developers**: This mailing list is useful for people who have deep Snort code-level knowledge and have questions or discussions.

- **Snort OpenAppId**: This mailing list is useful for users who are interested in topics related to creating, troubleshooting, and maintaining OpenAppId modules and packages.

More details about these groups can be found at `www.snort.org/community`.

In addition, the Snort blog that is available at `blog.snort.org` is a valuable resource. It has quite a lot of articles related to Snort.

Summary

In this chapter, we discussed a variety of practical topics, such as migration challenges and operational challenges. We also discussed some of the common questions and concerns where we may need to do some debugging and troubleshooting of Snort. It is infeasible to list and discuss all the problems that can occur with a complex system such as Snort. This is where the Snort community is useful. We discussed the various community mailing lists that may be of help for someone who is trying to learn about Snort.

As we conclude our discussion and exploration of Snort 3, we have covered quite a lot of modules, functionalities, configurations and experiments. The aim of these chapters was to introduce Snort 3 to the reader, to enable them to configure the system to their needs, and to enable them to create custom signatures to solve their security problems. We sincerely hope that this book has provided you with sufficient information to get a head start in using Snort 3 for their specific security needs.

Index

packtpub.com

Subscribe to our online digital library for full access to over 7,000 books and videos, as well as industry leading tools to help you plan your personal development and advance your career. For more information, please visit our website.

Why subscribe?

- Spend less time learning and more time coding with practical eBooks and Videos from over 4,000 industry professionals

- Improve your learning with Skill Plans built especially for you

- Get a free eBook or video every month

- Fully searchable for easy access to vital information

- Copy and paste, print, and bookmark content

Did you know that Packt offers eBook versions of every book published, with PDF and ePub files available? You can upgrade to the eBook version at packtpub.com and as a print book customer, you are entitled to a discount on the eBook copy. Get in touch with us at customercare@packtpub.com for more details.

At www.packtpub.com, you can also read a collection of free technical articles, sign up for a range of free newsletters, and receive exclusive discounts and offers on Packt books and eBooks.

Other Books You May Enjoy

If you enjoyed this book, you may be interested in these other books by Packt:

Effective Threat Investigation for SOC Analysts

Mostafa Yahia

ISBN: 978-1-83763-478-1

- Get familiarized with and investigate various threat types and attacker techniques
- Analyze email security solution logs and understand email flow and headers
- Practically investigate various Windows threats and attacks
- Analyze web proxy logs to investigate C&C communication attributes
- Leverage WAF and FW logs and CTI to investigate various cyber attacks

Security Monitoring with Wazuh

Rajneesh Gupta

ISBN: 978-1-83763-215-2

- Find out how to set up an intrusion detection system with Wazuh

- Get to grips with setting up a file integrity monitoring system

- Deploy Malware Information Sharing Platform (MISP) for threat intelligence automation to detect indicators of compromise (IOCs)

- Explore ways to integrate Shuffle, TheHive, and Cortex to set up security automation

- Apply Wazuh and other open source tools to address your organization's specific needs

- Integrate Osquery with Wazuh to conduct threat hunting

Packt is searching for authors like you

If you're interested in becoming an author for Packt, please visit `authors.packtpub.com` and apply today. We have worked with thousands of developers and tech professionals, just like you, to help them share their insight with the global tech community. You can make a general application, apply for a specific hot topic that we are recruiting an author for, or submit your own idea.

Share Your Thoughts

Now you've finished *IDS and IPS with Snort 3*, we'd love to hear your thoughts! Scan the QR code below to go straight to the Amazon review page for this book and share your feedback or leave a review on the site that you purchased it from.

`https://packt.link/r/1-800-56616-6`

Your review is important to us and the tech community and will help us make sure we're delivering excellent quality content.

Download a free PDF copy of this book

Thanks for purchasing this book!

Do you like to read on the go but are unable to carry your print books everywhere?

Is your eBook purchase not compatible with the device of your choice?

Don't worry, now with every Packt book you get a DRM-free PDF version of that book at no cost.

Read anywhere, any place, on any device. Search, copy, and paste code from your favorite technical books directly into your application.

The perks don't stop there, you can get exclusive access to discounts, newsletters, and great free content in your inbox daily

Follow these simple steps to get the benefits:

1. Scan the QR code or visit the link below

https://packt.link/free-ebook/9781800566163

2. Submit your proof of purchase
3. That's it! We'll send your free PDF and other benefits to your email directly

www.ingramcontent.com/pod-product-compliance
Lightning Source LLC
LaVergne TN
LVHW081521050326
832903LV00025B/1572